The 4 States of Consciousness

Michael Mathiesen

"The most incomprehensible thing about the universe is that it is comprehensible." - Albert Einstein

By Albert Einstein

"I want to know how God created this world. I am not interested in this or that phenomenon, in the spectrum of this or that element. I want to know His thoughts; the rest are details.

Science without religion is lame. Religion without science is blind.

My religion consists of a humble admiration of the illimitable superior spirit who reveals himself in the slight details we are able to perceive with our frail and feeble mind.

The further the spiritual evolution of mankind advances, the more certain it seems to me that the path to genuine religiosity does not lie through the fear of life, and the fear of death, and blind faith, but through striving after rational knowledge.

Everyone who is seriously involved in the pursuit of science becomes convinced that a spirit is manifest in the laws of the Universe-a spirit vastly superior to that of man, and one in the face of which we with our modest powers must feel humble.

The scientists' religious feeling takes the form of a rapturous amazement at the harmony of natural law, which reveals an intelligence of such superiority that, compared with it, all the systematic thinking and acting of human beings is an utterly insignificant reflection.

There is no logical way to the discovery of elemental laws. There is only the way of intuition, which is helped by a feeling for the order lying behind the appearance.

The intuitive mind is a sacred gift and the rational mind is a faithful servant. We have created a society that honors the servant and has forgotten the gift.

The most beautiful thing we can experience is the mysterious; It is the source of all true art and science.

We should take care not to make the intellect our god; it has, of course, powerful muscles, but no personality.

Whoever undertakes to set himself up as a judge of Truth and Knowledge is shipwrecked by the laughter of the Gods.

When the solution is simple, God is answering.

God does not play dice with the universe.

God is subtle but he is not malicious.

A human being is a part of the whole, called by us Universe, a part limited in time and space. He experiences himself, his thoughts and feelings as something separated from the rest-a kind of optical delusion of his consciousness. This delusion is a kind of prison, restricting us to our personal desires and to affection for a few persons nearest to us. Our task must be to free from this prison by widening our circle of compassion to embrace all living creatures and the whole nature in its beauty.

Nothing will benefit human health and increase the chances for survival of life on Earth as much as the evolution to a vegetarian diet.

The man who regards his own life and that of his fellow creatures as meaningless is not merely unfortunate but almost disqualified for life.

The human mind is not capable of grasping the Universe. We are like a little child entering a huge library. The walls are covered to the ceilings with books in many different tongues. The child knows that someone must have written these books. It does not know who or how. It does not understand the languages in which they are written. But the child notes a definite plan in the arrangement of the books—-a mysterious order which it does not comprehend, but only dimly suspects.

The important thing is not to stop questioning. Curiosity has its own reason for existing. One cannot help but be in awe when he contemplates the mysteries of eternity, of life, of the marvelous structure of reality. It is enough if one tries merely to comprehend a little of this mystery every day. Never lose a holy curiosity.

What I see in Nature is a magnificent structure that we can comprehend only very imperfectly, and that must fill a thinking person with a feeling of humility. This is a genuinely religious feeling that has nothing to do with mysticism.

The finest emotion of which we are capable is the mystic emotion. Herein lies the germ of all art and all true science. Anyone to whom this feeling is alien, who is no longer capable of wonderment and lives in a state of fear is a dead man. To know that what is impenetrable for us really exists and manifests itself as the highest wisdom and the most radiant beauty, whose gross forms alone are intelligible to our poor faculties - this knowledge, this feeling ... that is the core of the true religious sentiment. In this sense, and in this sense alone, I rank myself among profoundly religious men.

The real problem is in the hearts and minds of men. It is easier to denature plutonium than to denature the evil spirit of man.

True religion is real living; living with all one's soul, with all one's goodness and righteousness.

Intelligence makes clear to us the interrelationship of means and ends. But mere thinking cannot give us a sense of the ultimate and fundamental ends. To make clear these fundamental ends and valuations and to set them fast in the emotional life of the individual, seems to me precisely the most important function which religion has to form in the social life of man.

When forced to summarize the general theory of relativity in one sentence: Time and space and gravitation have no separate existence from matter. ... ***Physical objects are not in space, but these objects are spatially extended.*** In this way the concept 'empty space' loses its meaning. ... The particle can only appear as a limited region in space in which the field strength or the energy density are particularly high.

The religion of the future will be a cosmic religion. It should transcend personal God and avoid dogma and theology. Covering both the natural and the spiritual, it should be based on a religious sense arising from the experience of all things natural and spiritual as a meaningful unity.

We cannot solve our problems with the same thinking we used when we created them. Anyone who has never made a mistake has never tried anything new.

The most beautiful experience we can have is the mysterious. It is the fundamental emotion that stands at the cradle of true art and true science. Whoever does not know it and can no longer wonder, no longer marvel, is as good as dead, and his eyes are dimmed.

Everything is determined, the beginning as well as the end, by forces over which we have no control. It is determined for the insect, as well as for the star. Human beings, vegetables, or cosmic dust, we all dance to a mysterious tune, intoned in the distance by an invisible piper.

"All this harmony (in the universe) could not have come into being without a supreme orchestrator.

No problem can be solved from the same level of consciousness that created it."

Einstein also famously said that if the Quantum Physics of his time was correct, then the world would be crazy. He was right, the world is crazy.

Einstein also could not foresee all of the recent discoveries in Quantum Physics and so he never knew that all of his projections of the spirit in the universe would someday be as proven as his great formula that defined the absolute nature of light and the role that it would play in determining the size and shape of our universe.

If he had been alive today to learn about the discovery of the God Particle using the Large Hadron Collider in Geneva, Switzerland and announced on July 4th, 2012, he would have rested knowing that everything he thought about the Creator of this magnificent universe and our role in it was true, as true as any physical manifestation of his discoveries in the form of modern day electronics and weapons.

He also would have known that his theory about the one single Force that is the precursor of every other force in the universe would have been proven as well.

I am both confident and hopeful that my observations here in this book may someday serve as a platform for future Einsteins to go forth and define in even more detail the workings of the Creator and the finely tuned and well-oiled machine that he made out of whole cloth for the sole purpose of feeding back to him the infinite beauty and amazing simplicity of it all.

In Einstein's words above, I also hope that you can detect the overriding sense of gratitude and humility that this great thinker experienced in his own consciousness. It is with this same sense of gratitude and humility that I give to the world, the basis of that great man's consciousness which is shared equally by one and all.

UPDATE:

When this book was first authored by myself, I had not known about some recent evidence brought about by people who study BACTERIA

It has now been shown that some species of Bacteria, probably the earliest form of life on this planet going back about 3 billion years when life first began, have the unique ability to digest electrons.

They are called Shewanella Oneidensis - shown here.

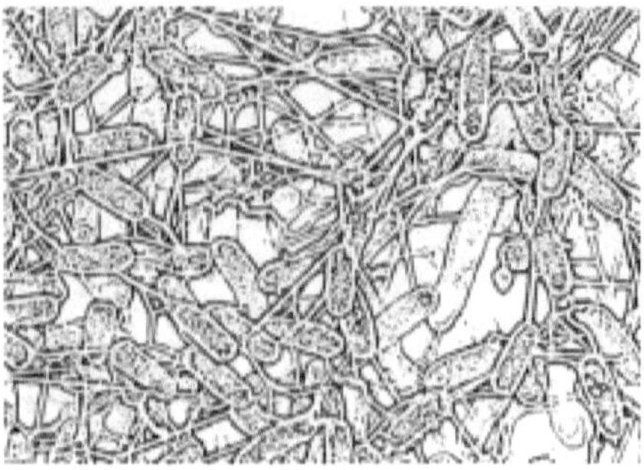

AND Geobacter shown here.

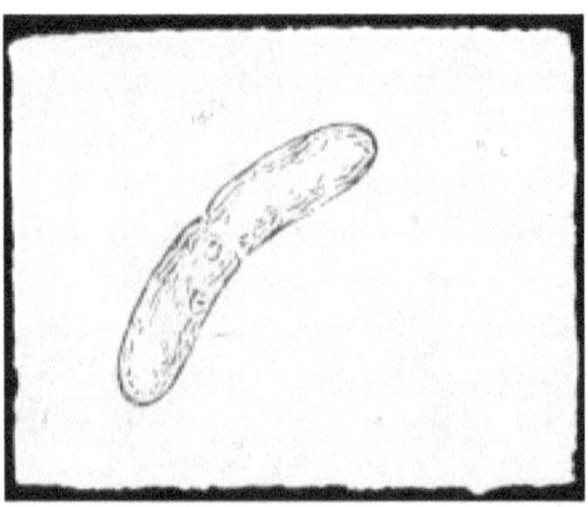

When you understand that more than HALF of our human DNA is exactly the same as Bacterial and Virus DNA, and that they can eat and digest ELECTRONS, I take that as proof that our own DNA, descended from Bacterial DNA is the basis of our own State of Consciousness that we have gained in the 4 states of Consciousness that I explain in this book.

As you will learn, and I theorized in this book that the very essence of your being - your CONSCIOUSNESS didn't just arrive on the scene the day you were born, there were billions of years of Evolution that had to take place and that goes into the chemical mix of what you are, what you do, how you think and even the major events of your life.

This evidence above of Bacteria that actually use ELECTRONS to fuel there existence on planet Earth instead of food, combined with the evidence I have given here in this book is PROOF that we have achieved so much in our Evolution only because the subatomic world has allowed it all to happen and had the first THOUGHTS about us all.

Introduction

A Home Demonstration

Before we start talking about the latest discoveries in Quantum Physics, I thought it might be useful to first provide a simple demonstration of these concepts to the reader using the laws of physics that everyone who has turned to their microwave oven to prepare dinner or a snack in seconds already knows.

The microwave oven.

The 1st State of Consciousness - Electronic

When you first turn on your microwave oven, you are turning on the switch on the circuit in your home that allows the flow of electrons from your local electrical power plant down the line into your home and then finally into your appliances. In this case the new way of cooking things from the inside out within seconds instead of hours.

Whenever you flip on the lights or the hair dryer or the washing machine in your home, you are in fact utilizing the 1st state of Consciousness, the Electronic State. Don't worry, in the first chapter of this book, we will demonstrate how and why electrons exhibit the property of Consciousness. For now, I only want to show you how all states of Consciousness in this universe snaps together for the sole purpose of keeping you and I on a level playing field.

So, in the first state of Consciousness, we utilize our electrons to perform work for us and the microwave oven does this very easily and obviously for anyone who understands what the microwaves being fired out of this amazing new invention have done. First of all microwaves are radio waves, but they are very small and energetic and they are produced by the microwave oven from a very simple little device called a MAGNETRON - seen here.

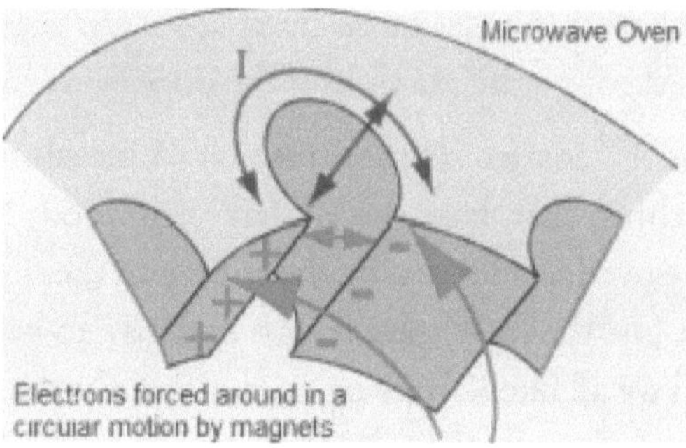

Inside this little device, invented in the 1950's by an engineer at Raytheon Electronics, you can see above how the electrons you have released along your house's wiring system into the oven when you turn it on. The genius of this little device is that it takes your electrons and sends them into these round chambers through the use of magnets. When the electrons swim around in these circular chambers they produce the radio waves, you know as microwaves - the blue lines.

The 2nd State of Consciousness - Rotational

This is a really good way to visualize in your mind's eye, the 2nd State of Consciousness - the Rotational. If we had not discovered how to send electrons round and round inside this little Magnetron, we would not be using the Rotational State of Consciousness to help cook our food and dramatically streamline our lives.

So far, in this part of our home demonstration of the 4 States of Consciousness, there is one more word that we need to learn - 'Induction'. Induction means simply to make something happen.

When, the Electronic State of Consciousness enters this man-made device - the Magnetron, what is happening is that the 1st State of Consciousness - held by the electrons, and thus aptly named - the Electronic State - actually induces the 2nd State of Consciousness or the Rotational State. And, don't worry, we're going to clarify what it means for any part of Nature to be Conscious.

We can now trust in these terms because we can see them in use in our daily lives - through the example of the microwave oven.

The 3rd State of Consciousness - Vibrational

The next thing that happens is that your food, sitting on the turntable of the oven, is radiated by the energy of the microwaves. The microwaves are just the right wave length for them to heat up the food in the oven and we all know from experience that the more water in the food, the faster and evenly the food heats up. This is because the water molecule is a very simple one. Most of us can easily remember that water is simply two Hydrogen atoms and one Oxygen atom bonded together and the symbol is therefore - H_2O.

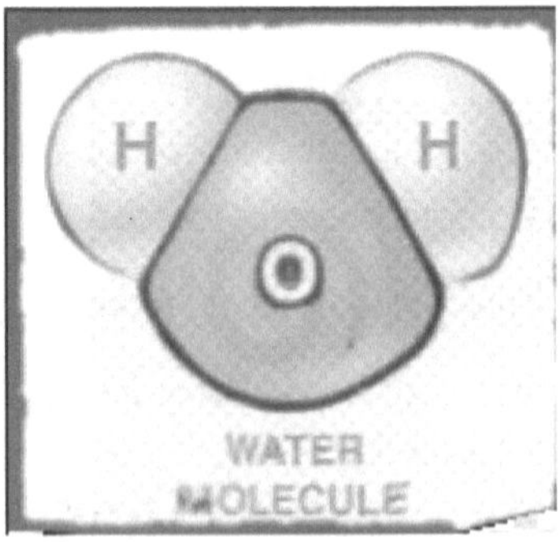

And, it's also important to note here that whenever atoms bond together in any combination, we now enter the land of molecules.

Therefore water is a molecule as is the substance of all the food that you will ever consume and even the cells in your body are composed of molecules as well. In fact, I believe it's safe to say that everything we are, happens on the molecular level. We only perceive the whole being of the trillions upon trillions of our molecules as one unit - the animal we call a human being.

But, in reality, we can better be described as a giant bag of molecules that have formed a community to give us an overall experience of the self.

So, in this process of induction of Consciousness states, we've gone from electrons, to waves of energy, and now into the molecules that make up our food and the molecules of the food we ingest are going to be absorbed by the molecules of our heart muscle, our brains, our fingers and toes, our eyes, our ears, our nose.

The Vibrational State of Consciousness

So, on the molecular level, we now have the Rotational State of Consciousness as it prepares our food molecules for ingestion. Inside of the molecules of any kind, we have induced a more energetic state of energy known as the Vibrational state.

All molecules vibrate because they contain energy. This is a kind of potential energy and we mostly know about this concept from the discovery of Albert Einstein's famous equation of E = mc2 - of course. In other words, the energy contained inside of any atom or molecule is equal to its mass multiplied by the speed of light squared. Since the speed of light - 180,000 miles per second is such a huge number, it makes the amount of energy held within any mass almost infinite and we all know that this knowledge of ours has led us to the brink of nuclear war that would spell the end of civilization as we know it.

But, up until the time that any molecule or element can be induced to explode in our faces, the molecule will contain this almost infinite energy in the form of 'Vibrations'. This is what I have termed the 3rd state of Consciousness or the 'Vibrational' State of Consciousness.

One of the most important laws of physics is the 'Conservation of Energy' rule and simply put means that energy is never created in our universe, it is only transferred or 'Conserved' from one state into another. Therefore, in the case of our microwave oven, the natural 'Vibrational' energy of our food molecules has been increased by the microwaves - The Rotational State - of our microwave oven. We can all agree that the energy state of our food has increased because it is hot. More succinctly, it has been heated by the interaction of the microwave energy and the molecular energy - friction - as the molecules vibrate more rapidly and this is perceived as heat by us.

The 4th State of Consciousness - The Feedback Loop

When we put our microwaved food into our mouth, we can feel that it is cooked or not cooked enough and if so, we now consume the food with its raised energy state. All the flavors are going to congeal on our tongues and we can now say that the food is delicious and allow it to flow down into our stomachs where it is digested by more specialized molecules in our gut. And here in our gut, the food that we just prepared by utilizing the other three states of Consciousness is now converted into the carbohydrates and proteins that our body then utilizes.

Whenever we digest our food, we are of course getting our very life-sustaining energies from that food. Without food and water, we all die, our internal molecular processes simply stop functioning without the energy contained in our food being converted into energy we can use in our daily lives. Even the thoughts you are having in your head right now are powered by the food that you consumed last night or earlier in the day. This is the most basic and most easily understood example of how the Vibrational energy in our food is transformed into all of our thoughts, words and deeds. The Vibrational energy inside our food simply induces the next and the highest state of Consciousness - us.

In our bodies, the process of photo-synthesis, the energy from the sun as it is converted by the plant life on our planet into simple carbohydrates - molecules we can digest, is reversed and the energy from the sun is now converted into simple sugars which is then used by our molecules in our hands, eyes, heart and brain tissues and put into real work. We perceive the universe around us and when we are well fed, we say that the universe is 'Good' and when we are not well-fed, we see the universe as 'Bad'.

In both circumstances, whether we are part of the 'Good' things going on or the 'Bad' things going on, we are giving 'Feedback' to the rest of the universe. Our feedback comes in the form of crying or laughing, coldly witnessing events, taking joy in our day's events, or feeling the sorrows of them. In all areas of our existence upon this world, we are merely giving 'Feedback' to the other forces in the universe and in many cases we even ask the universe for a little 'Help'.

This is why I call the 4th and highest state of Consciousness - The Feedback Loop. We, as living things were toggled together to perform the work of giving feedback to the Creative Force - the 'Master Consciousness' of everything that the Creative force has done thus far.

But, the highest state of Consciousness is not reserved only for humans. All animals and all plants, this includes even the lowest forms of life such as bacteria and viruses, are all part of the Feedback Loop. Every form of life, no matter where it is found, no matter when it is found in any other part of the universe, is part of the 'Feedback Loop' because if this concept of mine holds, it must hold for all parts of the universe, not just here on the Earth.

In other words, the Creative Force that made this amazing universe of ours constructed it in such a way that it would be able to learn about the effects that the laws of Physics are having upon all living things no matter how vast and how far that universe may extend.

Also, in order for the universe to be uniform in all directions, the 'Feedback' must be instantaneous and be part of a constant flow.

Another important consideration is that at each step in the induction process, the level of energy at every step is amplified by a factor of roughly TEN THOUSAND TIMES. In other words, the energy state of the electron when it induces the rotational energy of our microwave oven is raised by about ten thousand times in power in order to create the microwaves that cook our food. At the next threshold of the Rotational inducing the Vibrational, the energy state is kicked up by another ten thousand times. So, the Vibrational energy of our molecules is now immense and even measurable in terms of the calories that we use every day.

Then, finally, at the fourth state of Consciousness, the energy is amplified once again at another ten thousand times the Vibrational Energy into our own brains and transformed into the fourth and highest state of Consciousness. We can appreciate this amplification process by observing the tiny impacts that electrons on their own can assert.

While at the other end of the Consciousness spectrum the human mind is able to build instruments to actually see and measure all of these great miracles of our universe and then even to use them in things like microwave ovens and nuclear weapons and Hubble telescopes and the Internet. If there were not an amplification process taking place between all of the states of Consciousness, we would never be able to accomplish even a fraction of what we have accomplished thus far.

Next, imagine what is possible in the future.

And lastly, as part of this simple 'Home Demonstration' of principles, it is important to note that whenever an extant organism discovers its unique and most important role in the universe as a 'Feedback mechanism', it must, by default enter a hyper-sonic acceleration of its evolution because it would of necessity desire to become more and more accurate.

What you will Learn

It all starts at the subatomic level, the level of reality that we notice every day. It starts out there, but cannot be contained at this level of smallness, and so must boot itself up to larger and larger things. This boot-up of the entire universe ends with us, the highest form of Consciousness, but this also means that we must feedback what we have observed and thus become the 'Feedback Loop', the end of the cycle, the ultimate goal of Consciousness, to see itself reflected back by the mirror of that which it creates.

And first, we must define Consciousness so that we are clear on everything else I have to say about it later in this book. Consciousness is simply a unit of existence that has become aware of itself. Nothing more, nothing less.

And since we now know that this awareness begins with the electron and that the electron carries an electric charge that we also use every day in our lives to power all of our machines as well as our brains, it appears to this writer that Consciousness and Electricity are one in the same. In living things, we can only use tiny amounts, millionths of electron volts. Whereas, the rest of the universe constantly exerting the power to form all of the stars, planets, galaxies, even Black Holes, must use much larger amounts of Consciousness, so many trillions upon trillions of volts that this outward curving section of the expansion of the universe must be measured in uncountable Giga-tons of pressure, the mind of God in action.

In a new discovery just a few years ago, Quantum Physicists learned that if they separated two electrons that had been paired, each electron would remember the state of spin of the other partner. When they changed the energy state of one - the other would immediately react and change to be back in harmony with its partner.

They dubbed this strange phenomenon - 'Entanglements'.

One electron spins UP and the other spins DOWN. The two are forever entangled in the knowledge of the other. Even separated for thousands of miles, when we change the UP spin of one of the partnered electrons, the other one, the DOWN spinning one, will change instantly to the opposite spin once again. There is no time delay and distance makes no difference to their knowledge of each other.

This is the first indication of 'Consciousness' in the universe and therefore since electrons are part of everything else that we know and trust, the 'Consciousness' of the subatomic world is passed all the way up through all other states. Most importantly, we demonstrate in this book precisely how this simplest and most basic form of Consciousness is passed on up the line of the physical world and into us, all members of the Feedback Loop - the 4th and highest state of Consciousness.

In the not-too-distant future, I predict that the same property of 'Entanglements' found in electrons will also be found in Quarks and even in particles below these because quarks also spin UP and others spin DOWN. Indeed 'Spin' appears to be a property possessed by every object in the universe.

I believe that no matter how many levels of Quantum Physics we uncover over the coming centuries, we will continually find that Consciousness will be the lowest common denominator and contained inside of every part of the universe that we uncover.

Please note that even though Quantum Physics may appear to be one of the most challenging of all the Sciences - when viewed through the prism of this author's eyes and hopefully also yours when you have finished this book - all the concepts contained herein are truly the easiest thing in the world to comprehend.

# # #

There are four and only four states of Consciousness and the state of Consciousness that you are using right now to read these words is part of the 4th and highest state - what I call the Feedback Loop.

We are here because Consciousness can not exist without feedback from the opposite end of the cycle. Thus, all of life eventually becomes the necessary goal and Consciousness found a way to get us here in the condition we're in today.

The latest and most current scientific discoveries in the areas of human Evolution by the Human Genome Project, in Quantum Physics with the discovery of the God Particle and in Astronomy from the discovery of thousands of other planets in the nearby universe who may harbor life - all comes together in this book about how and WHY the universe was created the way it is - basically a giant machine invented to create life and get feedback from all forms of life everywhere.

The four states of consciousness are:

- The Electronic - It all starts with the electron.
- The Rotational - Electrons band together to create molecules.
- The Vibrational - Molecules in turn communicate to create US.
- The FeedBack Loop - Essentially all life in the universe including you and me.

As the 4th and highest state of consciousness, we are invented in order to provide the Creator with information about how he or she is doing. As we give back this information to the universe, the Creator responds with higher and better challenges for us to overcome. Each challenge feedback increases the energy state of the universe and lowers our own - but as part of the feedback loop, eventually all of the energy we fed into the universe is given back to us with dividends and our lives improve greatly as our energy state is raised by an equal or higher amount to that which we emitted. It's really that simple. The greatest lesson you will learn from this course is that you must not ever overthink Consciousness.

It's really that simple.

# # #

New evidence of my theory has surfaced in the last few years. Researchers have now discovered something called 'The Zinc Spark.

At the point of conception of all mammals, including humans of course, there is a release of light from 15 - 18 billion Zinc atoms and all of it within about ten minutes after the sperm penetrates the egg and the human is created.

This is very exciting news because it validates the fact that the 1st State of Consciousness, the Electronic, is clearly introduced into our beings at this precise moment. The Electronic state of Consciousness, at this point in our lives induces the next state of Consciousness, the Rotational State. We now have photographs of the actual event.

The above is an actual photograph of the event when the Electronic State of Consciousness is transferred into the human animal, actually, actually into all mammals because all mammals derive from eggs seen above. (I believe all living things experience the 4 states of Consciousness, not just mammals. We have yet to have evidence of anything other than in mammals, however.)

The energy from the 'Zinc Spark' is transferred all around the egg and starts the 'Rotational State of Consciousness all within minutes from conception. The Vibrational State of Consciousness happens as the animal is old enough to perceive the universe around him or her via the five sense organs. Our resonating to the music of the universe in ways to be discussed, produces the third state of Consciousness, the Vibrational State. The Vibrational State of Consciousness, in the final stage of the closed loop of energy flowing around in the universe induces the 4th and final state of Consciousness - The Feedback Loop.

The Feedback Loop state of Consciousness is your awareness of this book, and all other events in your life as you report them back to the Source of your energy, the Creator. Whoever, or whatever the Creator is or where he hangs out is not the subject of this book.

We can infer that a Creator exists because the design of the universe as discussed here is a simple circuit through which all Consciousness must flow and the end result is a form of Consciousness which wants to give 'Feedback' to the universe almost everything it is experiencing. This is the 'Feedback Loop' and is the strongest evidence that some 'one' or some 'thing' invented the Laws of Physics for the sole purpose of creating this system of learning about what has been created by those who were created by these electronic circuits, and at the ultimate end of the process of Creation can even create circuitry that mimics those of the Creator.

In a follow-on book to this one, I explore the possibilities of using the energy that is produced by the Electronic State or - 'Maxtricity'. This book can be found at - 'Maxtricity.com'. It's an excellent follow up to the information you are gathering here.

# # #

In this book we will be re-defining what it means to be conscious and what it means to have Consciousness. Though that seems to be a very large and challenging concept for some of you, please remember that we will be making it all much simpler than any other reading that you've done on the subject and in fact, our motto or anthem for this course is that we 'Never over think Consciousness. It's that simple.'

As a kind of proof, I will tell you here that we're going to cover the deepest and most profound and complex new discoveries of the most important energies in the universe from the smallest subatomic forces to the largest forces out in space/time. Although that may seem daunting, you will see quite quickly from the very first chapter to the end, that these things are only complicated by a whole array of complex mathematical proofs for and by the scientific community that serve to gain confidence in their underlying ideas but which are highly unnecessary for the average person to grasp the most important aspects of how the universe is structured and why.

Albert Einstein once said that "The most incomprehensible thing about the universe is that it is comprehensible." In other words, it amazed Einstein that the workings of the universe are so simple that anyone can really understand the most important aspects of it. His major contribution was in showing us how simple space/time really is. It was his great genius in that he was able to prove some of them to the rest of us.

And also In this book, since the author claims no special brain power anywhere close to that of Albert Einstein, we cannot offer you the detailed proofs of these concepts, however, it will be important to note that all of the concepts have been proven already by others. And in this book, we only attempt to explain them all in a coherent and logical set of assumptions that puts all of our recent discoveries into their proper perspective.

All we ask is that the reader take each concept one at a time and try to connect these ideas as they are intended and in the Vibrational response that you may have upon the completion of this book, that you share these ideas with others because it is in the 4th and highest state of Consciousness, what I have coined as the 'Feedback Loop' that we achieve our highest energy states.

As soon as we can achieve this thorough understanding of your role in the Feedback Loop state of Consciousness and we make a habit of recognizing the connection that we all have to all other forms of Consciousness in the universe, all of our troubles will vanish, because there is nothing in the universe that we are not meant to be part of in the first place.

Chapter One

- The Electronic State of Consciousness

Consciousness, as far as we know today, begins at the level of the Electron. Electrons can know about one another. Someday, I believe this same awareness of one another will be found in the deeper level of the subatomic particles in the nucleus of the atom, the quarks, the gluons, the bosons, etc. Because everything we see in the universe contains a spin of one kind or another. It is the spin from within that comprises the condition that we're in.

For more details on the ideas in this book, please see my earlier work - The God Particle Bible. The following is the present state of Discovery of these truths from our latest Scientific Discoveries. Nothing of what you are about to read are from fables, urban-legend, fairy tales or children's bed-time stories. They are all the best conclusions we can make from the latest experiments involving the smallest things we know about in the universe all the way up to the biggest.

There are four and only four states of Consciousness and the state of Consciousness that you are using right now to read these words is part of the 4th and highest state - what I call the Feedback Loop. We are here because Consciousness can not exist without feedback from the opposite end of the cycle. Thus, all of life eventually becomes the necessary goal and Consciousness found a way to get us here in the condition we're in today.

The latest and most current scientific discoveries in the areas of human Evolution by the Human Genome Project, in Quantum Physics with the discovery of the God Particle and in Astronomy from the discovery of thousands of other planets in the nearby universe who may harbor life - all comes together in this book about how and WHY the universe was created the way it is - basically a giant machine invented to create life and get feedback from all forms of life everywhere.

Once Again – because this bears repeating.

The four states of consciousness are:

- **The Electronic - It all starts with the electron. Since they can know the state and location of one another, this is a form of Consciousness.**

- **The Rotational – Electrons, protons, etc. bunch up together to create molecules using the rotational energy within them.**

- **The Vibrational - Molecules in turn communicate with each other to create us and all other living things. Without molecules, we're dead.**

- **The Feedback Loop - Essentially all life in the universe including you and me, but now understanding our purpose**

As the 4th and highest state of consciousness, we are invented in order to provide the Creator with information about how he or she is doing. As we give back this information to the universe, the Creator responds with higher and better challenges for us to overcome. Each challenge feedback increases the energy state of the universe and lowers our own - but as part of the feedback loop, eventually all of the energy we fed into the universe is given back to us with dividends and our lives improve greatly as our energy state is raised by an equal or higher amount to that which we emitted. It's really that simple. The greatest lesson you will learn from this course is that you must not ever overthink Consciousness.

As I've said before, but which also bears repeating. It's really that simple. It's not Rocket Science.

# # #

In this book we will be re-defining what it means to be Conscious and what it means to have Consciousness. Though that seems to be a very large and challenging concept for some of you, please remember that we will be making it all much simpler than any other reading that you've done on the subject and in fact, our motto or anthem for this course is that we 'Never over think Consciousness. It's that simple.'

As a kind of proof, I will tell you here that we're going to cover the deepest and most profound and complex new discoveries of the most important energies in the universe from the smallest subatomic forces to the largest forces out in space/time. Although that may seem daunting, you will see quite quickly from the very first chapter to the end, that these things are only complicated by a whole array of complex mathematical proofs for and by the scientific community that serve to gain confidence in their underlying ideas but which are highly unnecessary for the average person to grasp the most important aspects of how the universe is structured and why.

Albert Einstein once said that "The most incomprehensible thing about the universe is that it is comprehensible." In other words, it amazed Einstein that the workings of the universe are so simple that anyone can really understand the most important aspects of it. His major contribution was in showing us how simple space/time really is. It was his great genius in that he was able to prove some of them to the rest of us.

All we ask is that the reader take each concept one at a time and try to connect these ideas as they are intended and in your responsive thinking that you may have upon the completion of this book, that you share these ideas with others because it is in the 4th and highest state of Consciousness, what I have coined as the 'Feedback Loop' that we achieve our highest energy states.

As soon as we can achieve this thorough understanding of your role in the Feedback Loop state of Consciousness and we make a habit of recognizing the connection that we all have to all other forms of Consciousness in the universe, all of our troubles will vanish, because there is nothing in the universe that we are not meant to be part of in the first place.

Consciousness, as far as we know today, begins at the level of the Electron. Electrons can know about one another. Someday, I believe this same awareness of one another will be found in the deeper level of the subatomic particles in the nucleus of the atom, the quarks, the gluons, the bosons, etc. Because everything we see in the universe contains a spin of one kind or another. It is the spin from within that comprises the condition that we're in.

It is the SPIN that we put on ideas that makes up our own egos, the true place where we live in our minds. Remove the EGO or the SPIN that we feel whenever we move about our world and there is only Consciousness that has been passed up to us from the lower states of existence, the molecular and even the atomic and subatomic states of swirling energies.

The God Particle Field creates all of the other particles in the universe including the electrons, protons, quarks, etc. The way that it does this is by putting a 'Spin' to an area in space time that spins so fast, it can act like a 'particle'. In reality the spinning is really just a wave. Waves of energy then define everything else that we know. There is nothing else there other than waves.

Whenever a proton - a positively charged particle - is created an electron will appear out of nothingness to counter the positive charge with its negative charge. The Electron - and presumably all subatomic particles by default will themselves into existence. This can only come from a higher force of Consciousness that gives them that will power. Thus the name for the Field that we have seen performing this task by using the most expensive Scientific Apparatus in history - The Large Hadron Collider and which is now known as the God Particle Field.

From Elementary Science proven over the centuries, the nucleus of the atom will also determine how many electrons it can have orbiting around it and how far away they can be, exhibiting yet another form of Consciousness. The Forces that we have determined to exist at this level of reality are merely what we have named with names like the Strong Force, the Weak Force, the Electro-Magnetic Force. These are merely more detailed ways of explaining the power of Consciousness and how it exists in the Scientific mind. In our minds, it's all evidence of a universal field of Consciousness that pervades everything.

The number of electrons in orbit around the nucleus of the atom will determine what type of element it will become. It is only the number of protons and electrons in partnership with each other that determines what role they will play in the formation of reality - based on their element's properties.

Everything in the universe is made out of a little over 100 elements. How do the elements maintain their elemental properties with the will to do so?

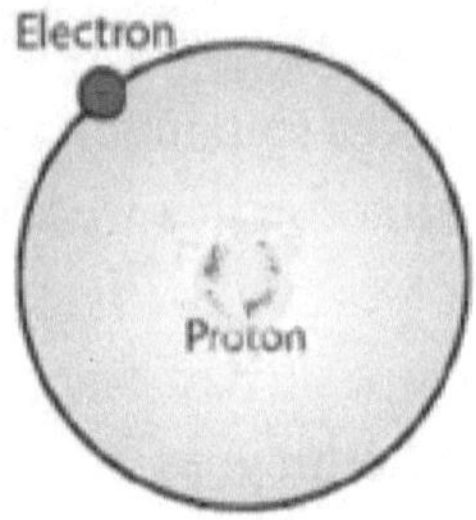

The electron doesn't orbit like the planets orbit around the sun. It orbits in what's called an orbital or a huge sphere. If the electron were the size of a planet and the nucleus was the size of the sun, the orbit of the electron would look like a huge cloud somewhere out in the orbit of Pluto that engulfs all the other planets of our solar system.

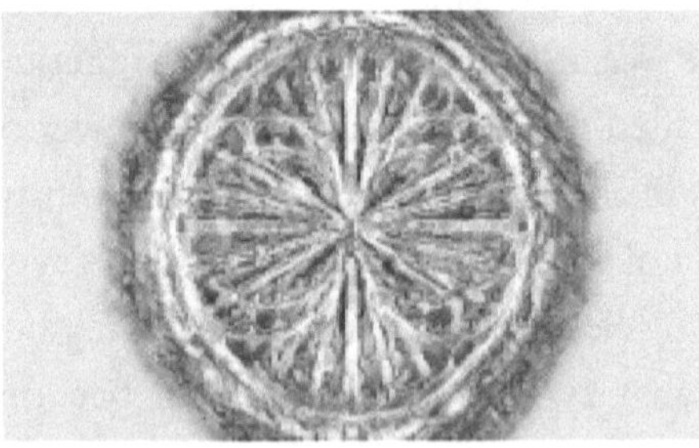

The electrons are the reason that we call the power that powers our homes and appliances as electricity. We were taught in Elementary School that electrons actually flow through the wires from the power plant to our home.

In reality it's only wave energy that is passed along the wires, there really are no particles being used up by our appliances. If that were so, we would be running out of electrons by now and have to find new ways to make them to send down the lines. But they do so in waveforms, not in particles as we like to imagine them, probably because it's easier to hold them in our minds in this way.

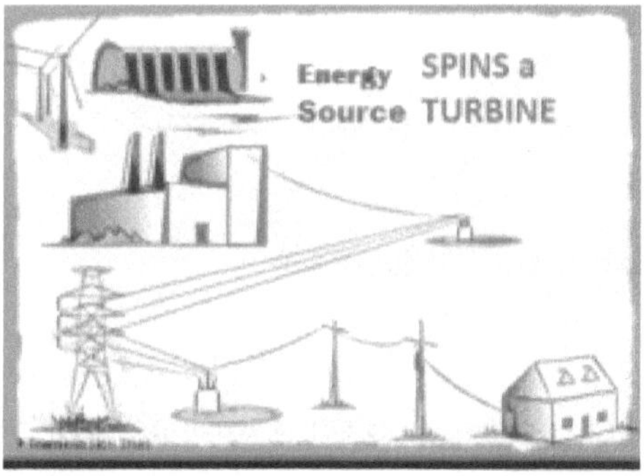

The power plant uses the energy found in water, oil or natural gas, the wind, coal or other sources to create steam which makes a turbine turn a generator.

A generator is a coil of copper wire that spins through a magnetic field and the wave energy caused by one source of energy and transformed into another is what we feel as 'Electricity'. It's actually just waves of energy flowing from the river, or the steam or the wind or the solar power into your home.

It's really just the energetic spin of things that creates electricity. And, on the Earth, it's the same thing - an energetic 'Spin' that created life on this planet.

It's all in the SPIN that we're in. Even our planet is SPINNING in Space. In fact, everything in the universe is spinning largely from the momentum of the spin of all of its smallest components spinning from deep within every cubic inch of the larger planets, stars, galaxies, etc.

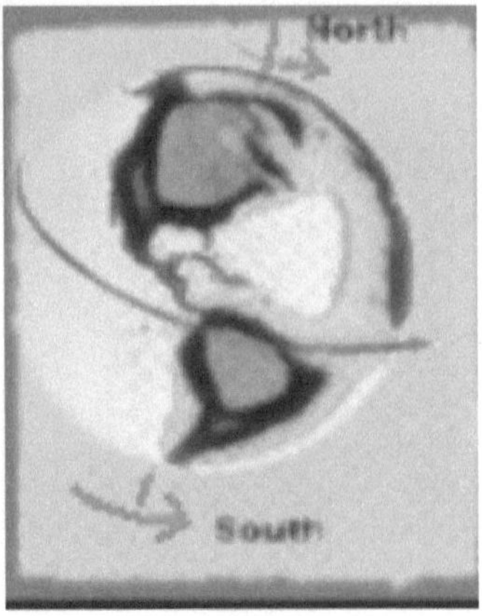

Our planet's SPIN is how we experience Day and Night. It has been spinning in space for billions of years and it spins through a magnetic field, which creates another form of Consciousness over time.

But as far as we know today, all of Consciousness begins at the level of the Electrons. Someday, we will probably find evidence of Consciousness in all of the activity of all particles even within the nucleus of the Atom. The Protons, the Quarks, the Fermions, the Preons and the Chameleons will all exhibit the property of Consciousness over all others.

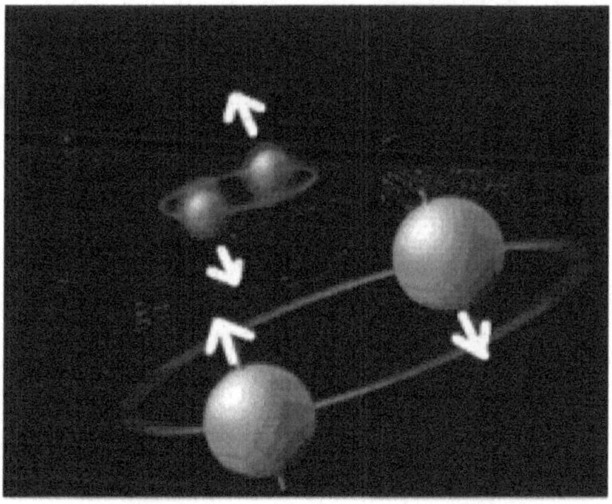

The best proof of these ideas comes from the concept recently discovered of 'Entanglements' - the UP Spinning Electron KNOWS the state of another electron it has been partnered with - HOW is that possible?

We know that they know about each other because when we reverse the spin of one of these paired electrons, the other one will change its state instantly to the opposite spin to be in sync with its partner once again. This takes place instantly no matter what the distance between the two. There is no time lost and there is no way that the two electrons can communicate with one another as far as we know.

How do they know the state of the other?

Consciousness is the only answer.

The electron is also bound to the nucleus of the atomic space that it occupies. This is called the 'Weak Force'. This is called 'Weak' only because there seems to be a stronger force within the nucleus that holds all of the nuclear particles of the atom together, the quarks that make up the proton for example.

There is also a set distance from the nucleus where the electron can maintain its orbital around the nucleus. It is true for every orbital from the closest ones to the most distant orbits the Electron must assume around the nucleus of the atom.

Every atom in the universe has this distance pre-set as well as how many electrons can fit into any Atom shell or orbital. There has never been found even one atom that does not maintain this set distance or density rule. How does the electron know where and or how to maintain their allotted distance and number and never vary? Or perhaps it is the nucleus of the atom that is the carrier or this knowledge of the distance. How does the nucleus know where its electrons are or how many?

Something, whether it is the nucleus of the atom, the protons, quarks, etc., or it could be the electron, but something in this relationship has to know where to place itself and it must be continually monitoring where it is in space so that it may never vary from this set distance. How does it know?

Consciousness of one's place in the universe as a relation to another is the only answer.

It is the combination of all the subatomic forces, plus gravity that keeps the universe from flying apart at any time and keeps us all together too.

Einstein believed that all of these forces could be part of only one overriding force. This is called the 'Unification Theory' and has yet to be proven. However, I believe that the work found here in this book is the final proof of this theory. Hopefully, you will be able to agree by the time you have finished this book.

Another of the most interesting new discoveries in Physics is what is known as the 'Laws of Probability'. When we try to observe the electron, we disturb it because we have to add or subtract energy from the atom in order to find out more about the electron. If we 'see' it, this means we must send at least a photon of light down into the atom which will add energy to the electron. When energy is added to an electron it is bumped up into the next highest energy state that must take place in a higher orbital.

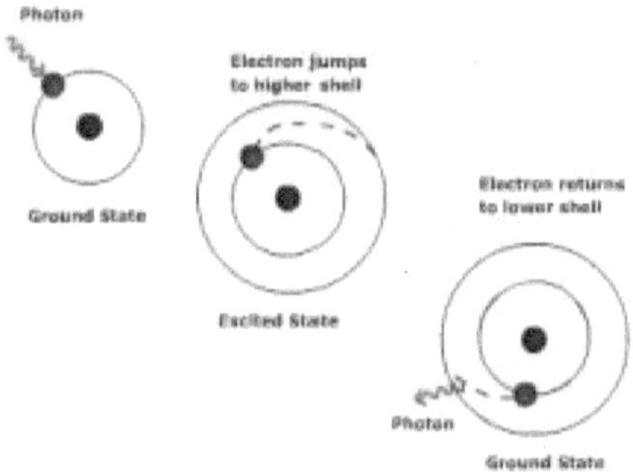

Conversely, when a photon of light is emitted from an electron, it must drop to the next lowest energy state found in the lower orbital or the ground state.

This is true of any system of Consciousness known to us because Consciousness itself is created at the atomic level and this is shown by the fact that whenever we who belong to the highest state of Consciousness observe anything in the universe we are emitting energy and therefore, we must drop to a lower energy state. When we receive feedback from the universe about what we have observed, we are gaining energy and therefore have our energy state expanded out to the next level.

An atom or an electron now wants to join with others of its kind. This is the nature of Consciousness. It must find partners to raise itself to the next state.

These atomic structures do this with the basic spin energy given to them by the God Particle Field.

The Electronic State of Consciousness then induces the next highest state of Consciousness - the Rotational.

The spin of the electron when combined with other atoms sheds some of its angular momentum with the rest of the structure. This is the essence of the molecular state of Consciousness - the Rotational State and we know this state as the material that makes up our skin, our brains, our eyes, our fingers and toes, etc. Because everything that we are made from is nothing more than simple molecular structures that have been made using the Rotational Energies of the Atom to produce structures such as proteins, tiny little mechanical structures that do things like transport sugar into your cells or transport the information within your cells to others, or join in the immune system to attack viruses and other diseases so that you can grow and remain a healthy individual.

When we say that one state of Consciousness induces the next level - it is our duty to explain the verb 'Induces', or the noun 'Induction'. This is nothing that we are unfamiliar with because every day, we see things due to the induction of the light waves all around us that are then reflected back into our eyes and gives us our eyesight.

Just in case the reader is not familiar with the concept of 'Induction', we can give the example of how light can travel from one end of the universe to the other without any other force to keep it moving forward than 'Induction'.

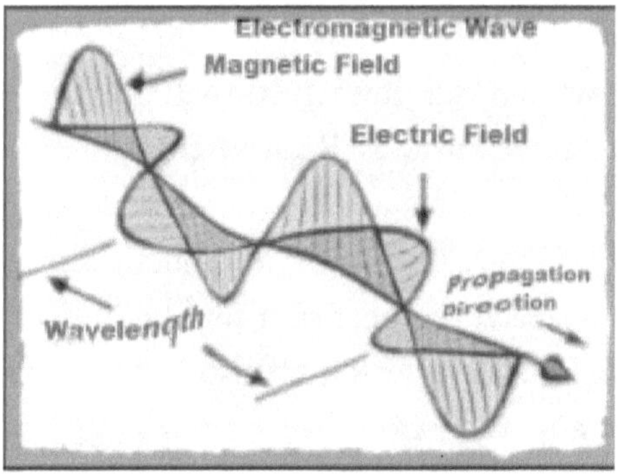

A beam of light can travel through space throughout all of eternity by simply inducing itself to exist in eternal travel. One wave, an electric field induces the electro-magnetic wave in the form of the light wave. The light wave, then induces another electric field, which in turn induces another light wave and on and on it goes for an infinite number of iterations

Now, wrap your mind around this thought experiment.

When you look up at the stars at night, you are witnessing light waves that left these stars millions and sometimes billions of years in the past. And, you can assume that intelligent life forms are looking up at the same stars from the opposite side of the universe. So, the light waves that you can see, are also inducing themselves to exist in a spherical shape that engulfs the universe in all dimensions, creating the dimension of Time as they travel these immense distances.

The light from distant stars radiates in all directions from the source of the light - the thermo-nuclear furnace of the star.

So, light waves are not only inducing themselves to exist in any single beam of light. They induce themselves to exist in all directions at once like the way that a soap bubble becomes bigger and bigger as you blow into it. There's a force we can call 'Induction' that has the power to create the next levels of what we know as 'Reality'. This force has yet to be defined.

However, simply put, this is the same power of induction that one state of Consciousness uses to reach the next state and because of this property, we can predict that Consciousness itself also travels in waves.

When we're feeling the most Conscious, we're actually feeling the wave energy of Consciousness.

Chapter Two

- The Rotational State of Consciousness

This section of this book will be found in greater detail from an earlier work of mine - The Origin of Creation.

The 2nd State of Consciousness is the Rotational State. Rotation is different from 'Spin'. So far, we've talked about the 'Spin' of the electron. This is what creates the concept known to us as 'mass' and this is the momentum that also gives the rotation to the over-all element that electrons and protons create when they combine . One momentum induces the next. It's like a spinning top that we played with as children.

Every state of Consciousness can be extremely subtle such as the molecular rotation in our bodies all the way out to the edges of the universe that are rotating right before our eyes. We can see that from our observations of the largest things we know that rotate - the galaxies.

Every galaxy in the universe rotates at varying speeds. Our own Milky Way Galaxy rotates at millions of miles per hour in space. This Rotation is induced from the uncounted billions of smaller particles that comprises all of the stars and planets of the Galaxy. We have given this force in the universe the name of 'Gravity', but what is Gravity? Has it ever been found? As of the writing of this book, no one has ever discovered the source of Gravity. The source of light is easy to discover. The source of what we call Gravity is much more obscure and it can be said now that Gravity stems from the forces inside of the Atoms that make up the larger objects that we have always believe have been the source of Gravity, but in reality, it is the smaller subatomic structures that make Gravity appear at the level that is much easier to observe.

At the center of most galaxies we've found to date is a Black Hole, which is also spinning on its axis or rotating. The deeper spins of the Rotational State of Consciousness is what produces even this massive spinning motion in space so powerful that it is felt billions of light years away.

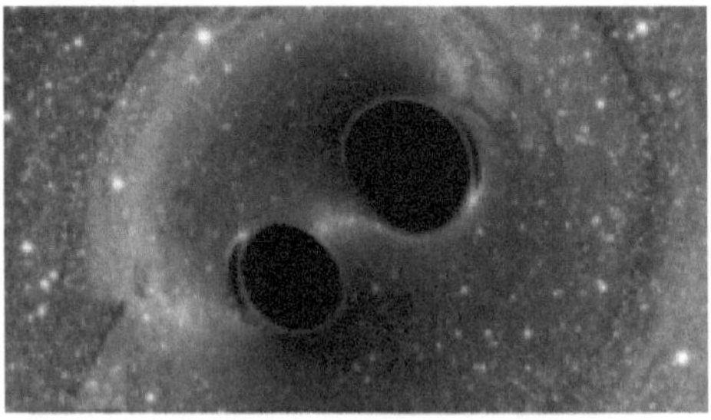

Black Holes have recently been found to emit massive gravity waves. Everything in the universe including Black Holes, the stars, the planets, all of the subatomic world is spinning in space. These massive Black Holes are more evidence that Gravity is really emanating from the smaller subatomic particle forces taken as a sum of them all.

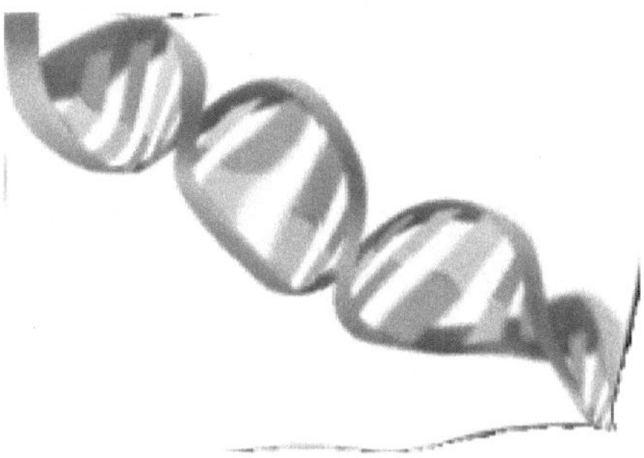

From the 2nd State of Consciousness - all life has evolved and continues to evolve. The Rotation is actually in our DNA molecule

You can clearly see the effect of the Rotational State of Consciousness of our own molecule which contains the code of life which determines every event in our lives. If this continuing inducement of the state of Consciousness to the next higher states did not exist, we could not exist, nor would any life in the universe be able to exist.

The 'Spin' of everything is how each state of Consciousness induces the next highest state. Each state of Consciousness has a higher value of Consciousness than the previous one since each has been induced by the lower value but must itself induce the next state.

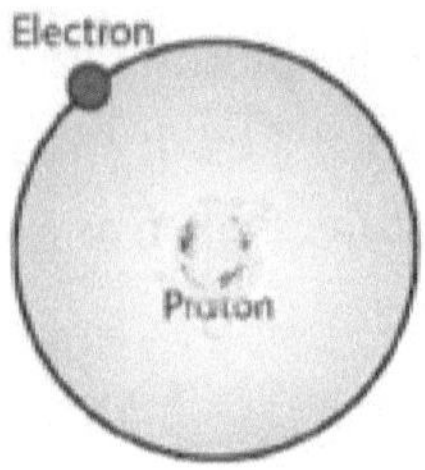

The spin of the Electron is a kind of momentum known as 'Angular Momentum' and this aspect of the Electronic state of Consciousness induces this next state of the Rotational simply by giving up some of this momentum to the next state.

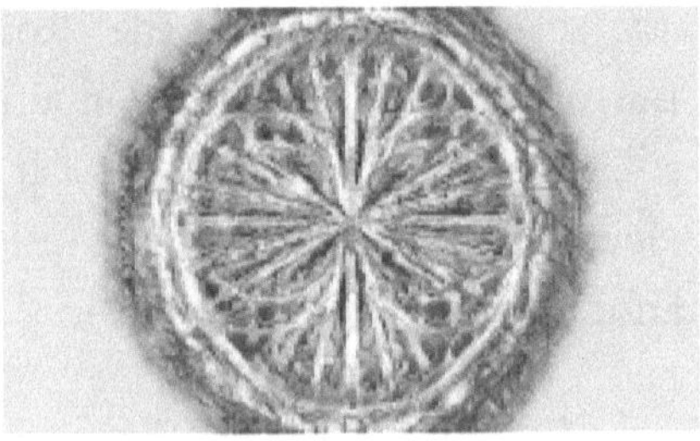

So, the electron is both spinning on its own axis and has either an UP spin to it or a DOWN spin. But, even while it is spinning freely in space, the spin momentum or angular momentum is inducing ROTATION.

This law of physics can be seen in the actions of the Spinning top we all played with as children.

Put the spinning top down on the floor and it will carve out this pattern of rotation on the floor every time.

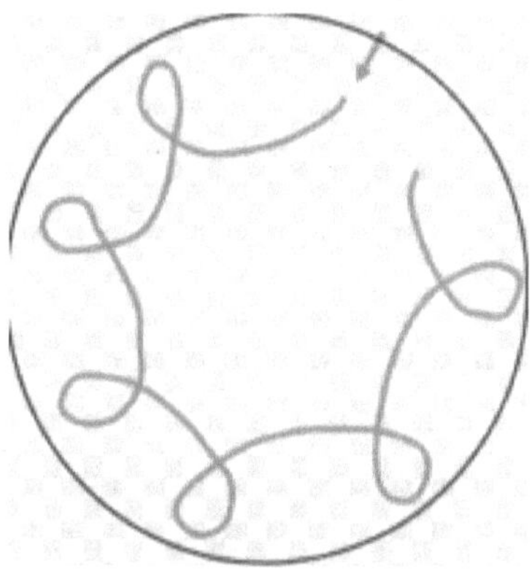

From the 2nd State of Consciousness - all life evolves because this is the first Molecular state of Consciousness. The DNA molecule was created in the earliest forms of life, bacteria, and co-exists in every cell of our bodies today and always will.

Some early forms of life, like this early Nautilus, even imitated the motion of rotation in their body morphology.

When one electron spins around the other inside the atom, it starts to create a wobble or a special kind of attraction of nearby units that the two elements then create as a combined unit of atoms. The Molecule is formed.

This creates the Rotational State of Consciousness because it is rotating along with everything else inside of it that is also spinning. The new element continues to carry the property of Consciousness because it is made up primarily of Electrons - which we now know - know about the state of the other. (Entanglements.) In the Rotational State, we now have two Atoms that are aware of each other and so they are now - bonded. Without this awareness of each other, there can be no molecular bonds.

The Rotational State of Consciousness resides in the molecules and heavy atoms, the elements that each atom has created in their constant and steady combinations. This is all demonstrated beautifully in what we know as the Periodic Table of the Elements. All of them, every element in the universe is made from this Rotational State of everything that is pasted together to make them. There are no exceptions that we know today.

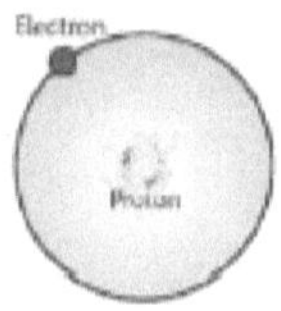

In the handing up of the Spin energy of the electronic state - the molecule absorbs this energy.

This absorbed energy induces the molecule to rotate. In absorbing energy a higher state is realized. When emitting energy the energy state of the object doing the emitting is lowered.

The entire essence of this book is really all about how either absorbing energy can raise your state of consciousness and emitting energy will lower it. It is that simple. There is nothing more profound about the universe other than this simple law of physics that is maintained from the smallest to the largest things known.

Atoms gain rotation as more and more spinning electrons are added, and then giving up this momentum to the molecule causes the Rotational State of Consciousness.

This means that every galaxy we can observe has had its energy state raised over billions of years to create this massive rotation of Consciousness. Does this mean that all the galaxies in the universe are aware of each other? Or that all galaxies in the universe are aware at least of their nearest neighbors? Since electrons carry this property of Consciousness - it's a safe bet that galaxies also have Consciousness of their own existence.

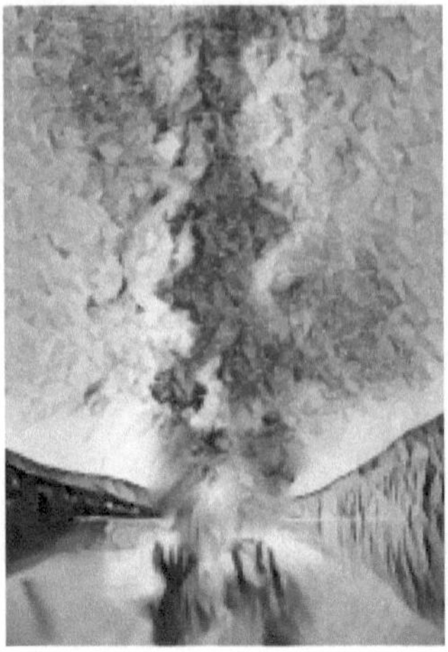

Whenever we look up at the night sky and observe the Milky Way Galaxy making its way across the sky, we are reminded of just how powerful the Rotational State of Consciousness can be.

Someday in the next billion years or so our own Milky Way Galaxy will collide with our nearest neighboring galaxy - the Andromeda Galaxy. There is somewhere an awareness of this partnering, just much more time involved than that of the paired electrons.

Recently Gravity Waves were discovered for the first time by monitoring two black holes - seen below - who are both spinning in place and revolving around each other some billions of light years away from us.

Even our DNA was produced by the Rotation of the Earth over billions of years. We can see it in the pattern of the Double Helical shape of our molecule which is basically the code of all life on our planet and probably coding the life forms found on every other habitable planet in the universe.

Indeed, it is my prediction that when a life form is found on any other planet - and I am hopeful that this will occur in my lifetime - that it will also have the same helical form and will even have some of the same coding. It's bound to have a completely different over-all code because life will need to adapt to completely different conditions on other worlds. However, I believe the DNA molecule will still be the basic architecture or blue-prints for any other life found in the universe, because it will have to follow the same laws of physics as we do here.

Imagine someday when an interstellar mission from the Earth is able to bring back a specimen of life from another planet. And, imagine if it is constructed out of the same DNA coding and imagine further - if you can - what implications that would have if scientists could then even splice some of the alien DNA onto our own or vice versa.

In fact, I believe whenever we can retrieve the code of life from another planet, we will begin to see new opportunities for new life forms through the joining of the two molecules or perhaps maybe just small splicings.

We may even get something like the aliens we project from our favorite Science Fiction.

The point is that Consciousness wants to become more and more conscious. It must become greater and greater in quantity and quality and therefore, one level of Consciousness wills itself to literally come out of its shell and claw its way up to the next level. This aspect of Consciousness is also replicated in how life continually moves to the next, more complex level of design over the millions of years of Evolution.

Everything in your life is coming to you due to the spin of something. It may be hard to imagine until you look back at your life and think about it as 'SPIN', the way that the media SPINS the news the way that they want you to hear it.

At the Rotational State of Consciousness it's like a dance. Two partners begin to be attracted to each other and move about the room, they are going to spin and rotate as much as they can and put all this motion together as an art form.

But at this point, we're not talking about any human behavior just yet. We're talking about the SPIN of a wave that kicks off all the music. There is no 'thing' that spins like the top. At the Electronic State of Consciousness, the first level that we know about, no object is actually spinning. There is only a wave of energy that appears to us to be in a state of a tight spin, making the probability that it exists somewhere, but it is just a probability of where it is. But, it leaves no doubt of 'what' it is or better stated, what role it will play in the universe.

In reality, particle energy is just there occupying space as an energy that we can measure in waves. If we were small enough to go visit an electron, there is no telling what it would actually look and feel like. It's my belief that this world would seem more like music to our ears, rather than something we would actually see or feel in any other way. But no one knows or probably will ever know that for certain since we probably will never be able to project our living selves into this realm. Anything is possible.

Who knows; there is no limitation to our inventive spirit or our ingenuity, it seems, and so perhaps someday, someone will invent a machine that can do this for us. At that point, we would be listening to the music composed by the most masterful artist of all time.

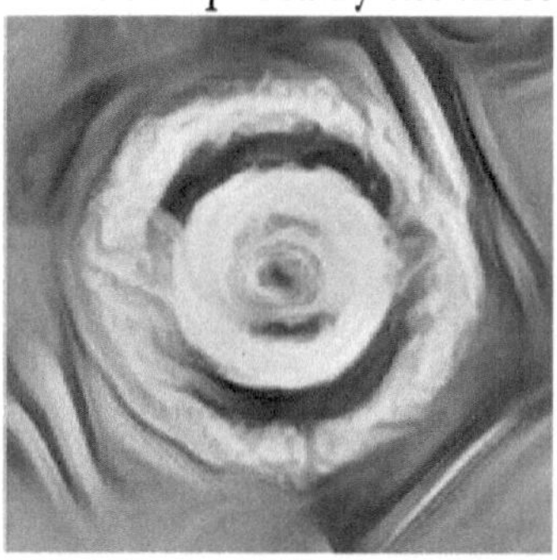

What we think of as particles are really just the flux between the top and bottom of the wave energies, the crests and the troughs. In this most basic and fundamental building block of our reality, things come and go. Nothing exists forever. The universe is merely twinkling on and off at such a rapid rate that it's too fast for our senses to catch this rapid flickering of things and so everything seems to be solid to us and lasting, when in reality, there is no such thing.

It may take billions of years, but we were created out of the wave motion of the Earth as it traveled through space over the last six billion years.

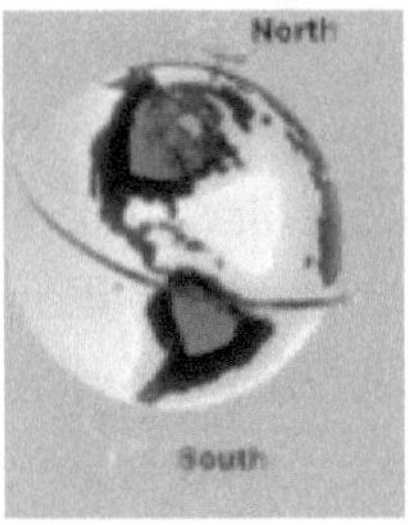

Our own lives are proof of how the Rotational state of Consciousness can then impact the next state of Consciousness, the Vibrational and then us - the Feedback Loop.

The Electronic State of Consciousness is responsible for the creation of molecules because the power of one electron wanting to find another and these two electrons wanting to find more groups and then all of this grouping forming the molecules that we all know and love - is what happens whenever the spin momentum or angular momentum of the electronic state collides with itself. This simple collision of subatomic galaxies in space creates the rotational momentum of all things from the atomic structure on up to the level of galaxies. Rotation observed anywhere in the universe is merely the manifestation of the same state of Consciousness - the Rotational.

The Rotational State of Consciousness allowed the DNA Molecule to take the form that it has taken. As the Earth rotated and spun around its axis

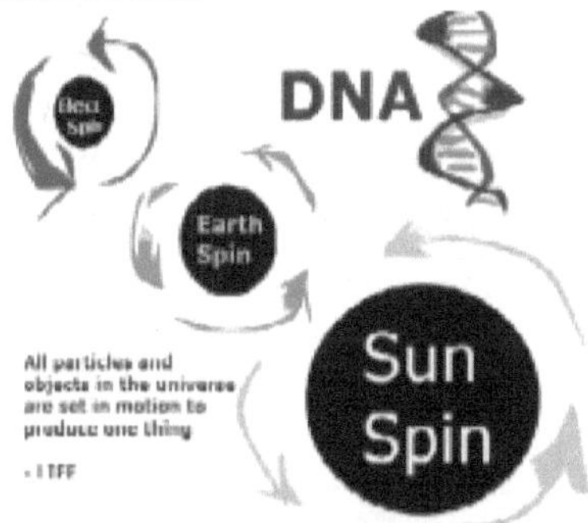

Could it be that the Spin that we're in here on Earth determined the Spin and Rotation of our basic Molecule of life - DNA?

Our DNA Molecule formed from the Rotational State of Consciousness as it begins to rotate on its own from the angular momentum given to it from the atomic rotation and spin. As it rotates, it creates the code of life that has designed the lives of bacteria and then right on up to reptiles, birds, fish, the primates and then even to us humans.

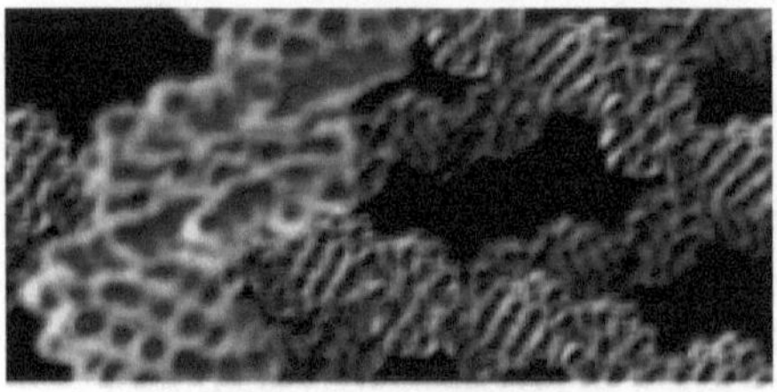

The shape of our DNA molecule replicates the description of the path that our Earth takes through the Heavens.

Whenever you look up at the night sky or look at the pictures produced from the amazing telescopes of today, you can know the power of the Rotational State of Consciousness. If this type of image doesn't vibrate inside you - you're not taking the time to utilize the power of your own Consciousness.

When I think about all of this - it seems to be truly the greatest miracle that life can evolve from something so simple as the mechanisms we are describing here in this book. The main lesson to be learned here is that you never want to overthink Consciousness.

Those who have not read this book will try to overthink it anyway, and many have done so over the centuries, but I believe those who have over-thought the process have really just confused us all more than enlightened us.

Chapter Three

- The Vibrational State of Consciousness

As we have seen, Consciousness, as far as we know today, begins at the Electronic Level, then works its way up the ladder to all of the next levels of Conscious awareness until it finally reaches us. Each level of Consciousness is simply induced by the previous wave energy in the way that every life form induces another more advanced species over the period of time known as Evolutionary time.

At the level of molecular forms in our universe, some of these basic organic molecules, such as proteins, start to vibrate in the presence of others. This vibrational response, like two tuning forks, to one another creates more and more complex molecules that eventually are able to do a job within our bodies. Our muscles, our cells, our blood is composed of molecules that can perform work. They would never be able to work together in harmony if it were not for the Vibrational State of Consciousness that is induced from the Rotational. In turn, the Vibrational induces the highest form of life - us - all living things - the Feedback Loop, which we cover in greater detail in the next chapter.

The way that simple forms of life induces other higher, more complex forms of life is through the waves of energy seen in our DNA molecule. As the waves are re-energized, the code held within is mutated and changed until something far more capable, bold and daring can come to be. This is how you and I came to be, through the sheer will of life to want to advance and enjoy the company of others.

We have just learned how the Rotational State of Consciousness induces the Vibrational State of Consciousness. Next step in the ladder of consciousness happens as the lower forms of consciousness simply will themselves up to more consciousness, more awareness of the universe. This is the concept of induction.

Everything in the universe comes to be what we see and feel as Reality in waves. There is nothing solid in the real sense of solidity because everything is really formed up in waves.

Below is a diagram of how an electron can be conceived. It's a double wave energy one up and one down. The red smudges in between the waves are what is perceived as particles in some experiments, but in reality, these areas within the waves come and go. Where the two wave patterns intersect, there are no more particles.

Waves inside the atom or waves that hit your ear from a guitar string are all the same - waves. You cannot hear, feel or touch an electron's wave, because the frequency is way out of our range of our eardrums. But, the sound waves created by a musical instrument can be heard by us.

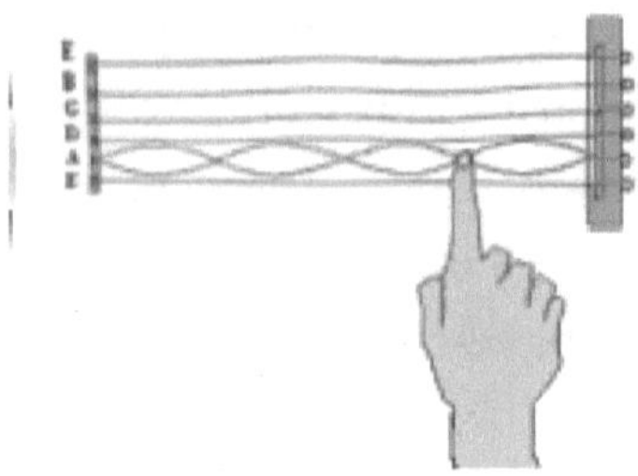

It's a sad truth, but we humans are only able to pick up on tiny fractions of the bandwidth of sound, light, all part of the symphony of music that is composed every day by the universe.

The Vibrational State of Consciousness is the bridge between the Electronic, the Rotational states and our own state - the Feedback Loop. The more we catch the waves of the Vibrational State, the easier and faster it is for humans to reach their maximum potential under the 4th State.

For me it was the Beatles music that taught me this great truth.

Whenever I listen to Beatles music I can feel my own energy state increasing rapidly and sharply to its highest it can be. I love all kinds of music, but for me, this music is the highest and best example of the art form.

Hopefully, you have your own favorite musician, artist, sculptor, painter, writer, poet, etc., because all art forms are the best example of the Vibrational State of Consciousness.

When we vibrate with a particular music that reaches our soul, or a painting, a book, a movie, etc., we are taking in the best energy that this artist could give us. Since we are then absorbing this energy, we are in a raised energy state of Consciousness. Whenever, we write great music and give it to the world, we are emitting energy and so are lowering our own energy state in an unselfish act of love for our world.

The reason that we are alive is because the Vibrational State of Consciousness persists as a gift from the Author of all things. Whenever our heart beats, for example, it is doing this only because the cells of this organ have learned how to act in unison and harmony and actually cause their individual actions as heart cells to co-operate into one smooth and useful motion that acts as a pump to circulate all of our blood to all sections of our body every minute of every day. Every cell in our body knows the role it is meant to play. This information is given to it by the DNA molecule sitting at the nucleus of every cell in our bodies.

If it weren't for the Vibrational State of Consciousness, we would have no thoughts because our brain cells would not know how to be brain cells. If it weren't for the Vibrational State of Consciousness, we would have no way to touch one another because the cells in our fingers would not know how to transmit the sensation of touch to our brains. . If it weren't for the Vibrational State of Consciousness, we would have no way to see one another because the cells that make up our eyes would have no way to know how to transmit the light waves that hit them to our brains for processing of our sight.

If it weren't for the Vibrational State of Consciousness, we would have no way to hear one another because the nerve cells in our ears would have no way to know how to convert the sound waves into an electrical signal that our brains can interpret as sound. . If it weren't for the Vibrational State of Consciousness, we would all die immediately.

Since our heart muscle is composed of electrons, atoms, molecules and all of the forces we've discussed so far, the Vibrational State of Consciousness, and all states below it can literally be seen and felt by anyone with a heart, every day.

The Vibrational state is literally the will of the Electronic and Rotational States of Consciousness to achieve something much greater than the mere sum of all its parts.

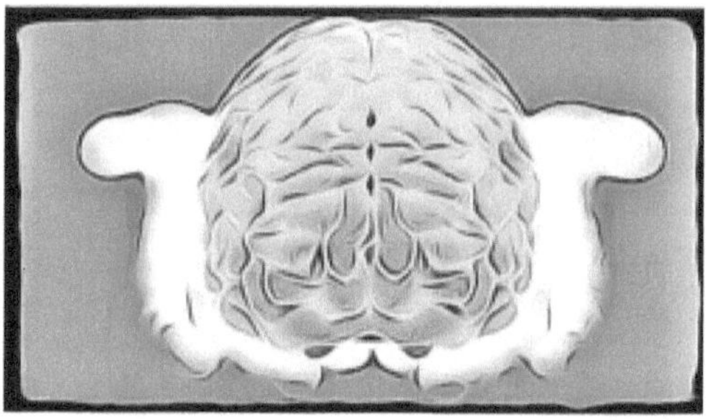

The brain is another excellent example of the structure of the electron coming out in a world of its own creation. It's the electro-chemical flow of brain waves that allows us our every thought and action throughout every day of our lives.

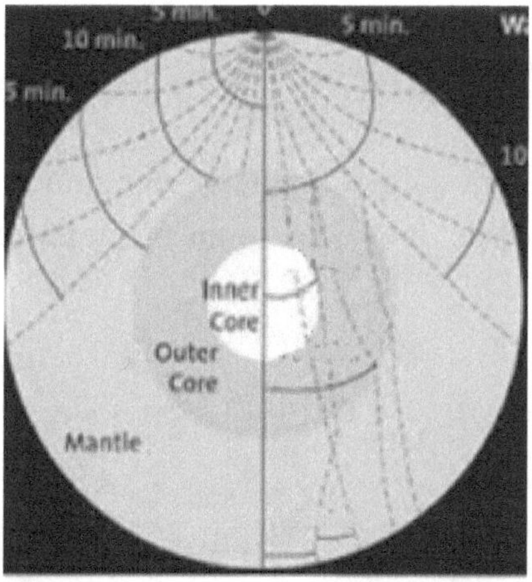

The Earth from its core on outward vibrates and sometimes enough to give us what we call 'Earthquakes' on the surface.

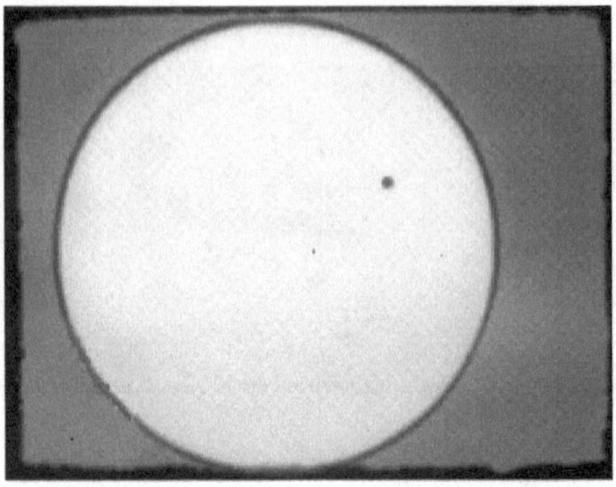

This view of the Sun shows the planet Mercury orbiting it in the top right.

The mighty thermonuclear reactor of the sun causes it to vibrate as well. Everything we know, if it has a consistent energy source vibrates in some frequency, most of which we need specialized instruments to detect.

It all starts with the vibrations of the subatomic particles, especially the electron bubble and moves on up the hierarchy of Consciousness.

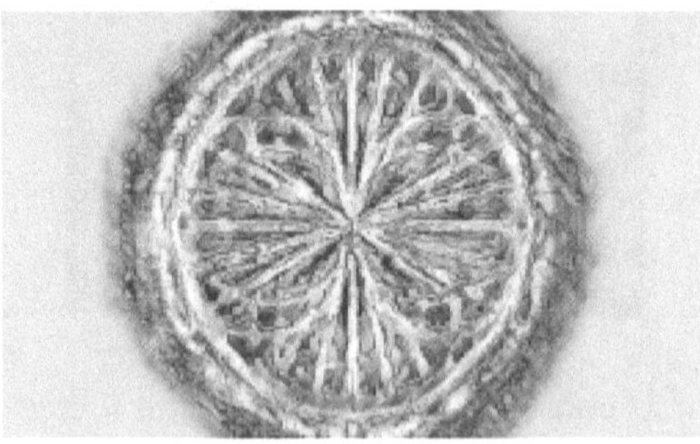

Thousands of trillions of billions of millions of these energy bubbles comprise our real world. As they move around each other and attach themselves to each other to make bigger and bigger things such as all of the molecules that we come to know as life forms, a vibrational energy, something completely independent of the formation itself will come into being as a form of the 'Expression of Consciousness'.

The human heart, for example, beats eighty to ninety times per minute for every human being for decades for most of us, keeping us alive, pumping the life-giving blood into every cell of our body because of the Vibrational state of Consciousness. Without it, we could not exist because the molecular interactions would be chaotic and unorganized enough so that no systems such as our breathing, digesting, growing, reproducing, even thinking could take place.

The Vibrational State of Consciousness is the music of the universe at the first level that we can actually feel it with our ears and love it with our minds. The Vibrational State is what happens after all of the events of the Rotational State. Nothing can vibrate with the first emanations of life until the Rotational State has produced all of the sub-systems of life itself, the breathing, the nervous system, the thinking brain, the muscle that gets us to where we need to go.

Another way to think about the Vibrational state of Consciousness is the way that flocks of birds can wheel and spin in amazing patterns with thousands of animals turning as one unit. There is a vibration of their lives that every individual in the flock or the school of fish can understand even without the advantage of language or any advanced technology that we know of.

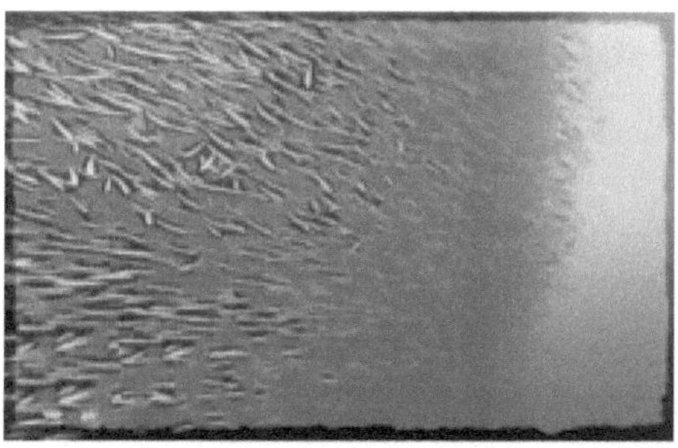

Almost all life forms that we know about seem to exhibit a joy of life that can only be from the Vibrational state of Consciousness where everything seems to be going your way.

If there were no waves in the oceans there would be no dolphins.

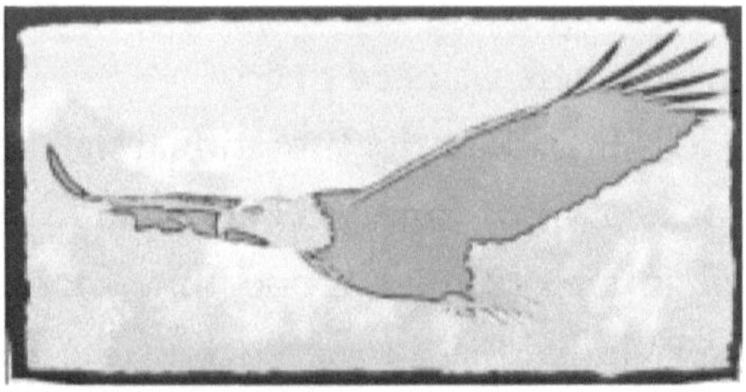

If there were no waves in the ocean of air above us - there would be no Eagles.

If there were no waves of Evolutionary changes there would be none of these noble creatures who gave birth to us. It is the Vibrational State of Consciousness that forced our descent from the earliest life forms to the more complex form of life that we possess today.

The good news is that you can experience the Vibrational State at any time you choose to do so. You've hopefully enjoyed the Vibrational state of Consciousness once at least in your life when you found that special chemistry we call love.

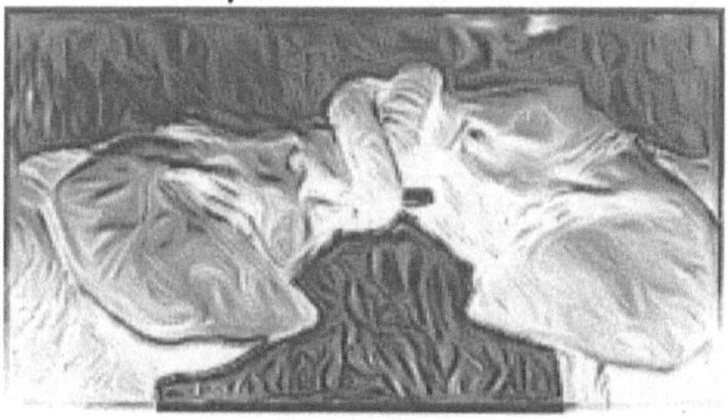

Elephants are known to remember dead family members. The Vibrational State mixing in with a little of the next highest state - the Feedback Loop.

The Vibrational state is responsible for the constantly flowing generations of life everywhere on this planet and everywhere in the universe.

Ducklings with the total acceptance of the vibrations of their mother. We can sympathize with the other animals with whom we share this planet. This is the Vibrational State of Consciousness inducing its way into our mind's eye.

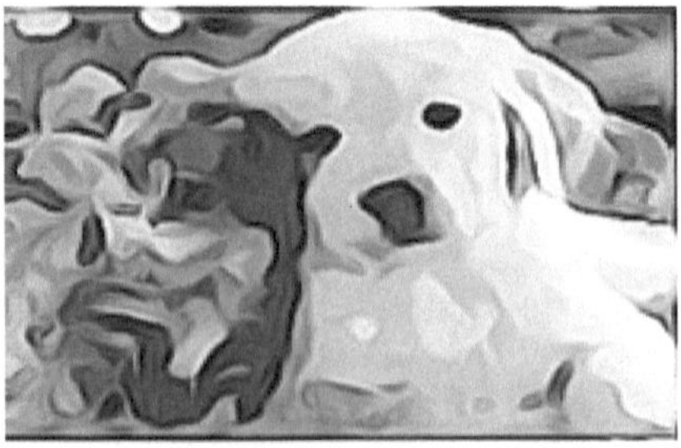

Dogs and cats living in harmony - total chaos? Or more evidence of the Vibrational state?

Psychologists have noticed for centuries that when people are put in charge of a pet such as a dog or a cat, their levels of stress are greatly reduced. Major diseases have been overcome. People generally live longer in the presence of a pet. Enjoyment of life is greatly enhanced. Some even call this form of relationship - the highest form of love they have known in their lives. What causes this? Could it be the Vibrational aspect of consciousness bumping up against each other?

When we're in love - most of us would say that we feel the vibrations of the hearts and souls of the other person in the relationship.

When swans take on a sexual partner, it's for life. If one of the partners dies, the surviving swan will never mate again. Do swans share the Vibrational State of Consciousness? Or is this just some basic instinct as we like to call it. What is instinct? Where does instinct come from?

The Vibrational State of Consciousness is all around us at all times and meant to be enjoyed by one and all at any point in our lives that we simply want to slow down and smell the roses.

And even the flowers themselves can feel the vibration of the Sun overhead and so they orient themselves to get the most out of the sun's energy every day by simply orienting toward the sun at every moment in the day.

Where do flowers contain an apparatus to learn the angle of the sun and where is the apparatus in the flower to turn its stem toward the sun at just the right angle at every minute as the sun moves across the sky?

All of life is possible because of the Vibrational state of Consciousness.

What got my attention one morning is when I suddenly realized as I was waking up that the path of the electron through space describes the pattern of our DNA molecule.

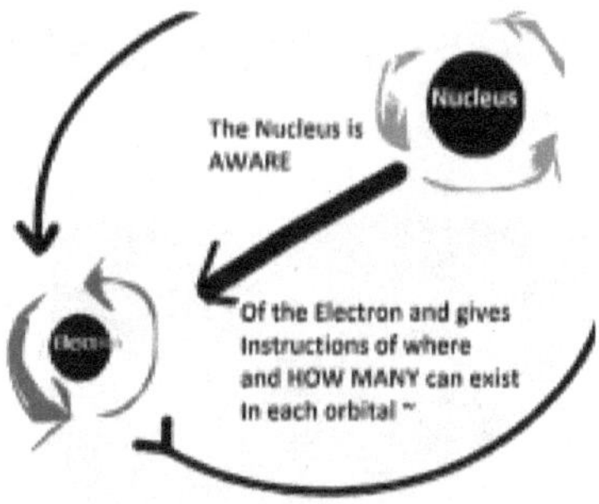
Nucleus
The Nucleus is
AWARE
Of the Electron and gives
Instructions of where
and HOW MANY can exist
In each orbital ~

As an electron spins in space it's also moving through space in a direct path. The two sides of the electron describe a double helical pattern, which by no coincidence is copied by our DNA Molecule.

Therefore the molecular realm of things is where the Vibrational state of Consciousness can exist.

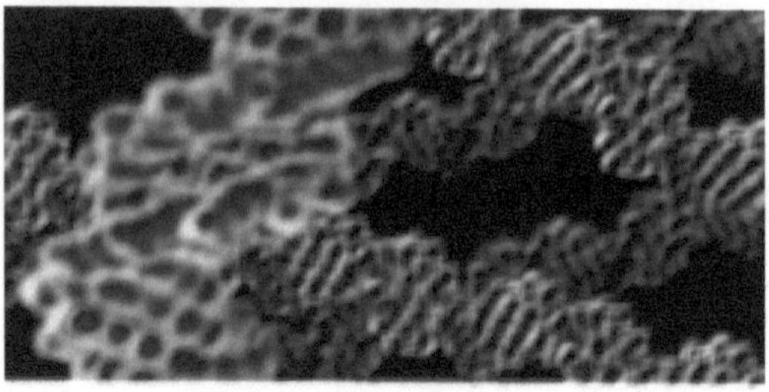

And, the Earth formed by the agglomeration of innumerable atoms and molecules over billions of years also forms the same helical pattern as it moves through space. Take two sides of the planet and trace them and you have the double-helix.

When Life is formed, it is from this Vibrational state of Consciousness and it even advertises its point of origin by painting an almost perfect picture of its author.

The shape of the DNA molecule is due to the earliest vibrations in the universe right on up to the formation of life on the Earth and probably every other life form found anywhere else.

All of life that we know shares our same molecule of the life code. It all started with bacteria who survived the early challenges of this planet with a poisonous atmosphere. Life is tough and it survived. In survival mode this early form of life transformed the poisonous atmosphere of this Earth, mostly carbon dioxide and hydrogen sulfide, basically what comes out of a diesel engine exhaust and turned it into the oxygen, nitrogen and helium that we enjoy breathing today. This process of the transformation of our planet took about one billion years, but in that time, the Vibrational aspect of our planet, a gigantic ball of molten lava, as it cooled and formed itself in space, continued to spike the DNA molecule with the information it would need to create more advanced and complex forms of life for billions more years.

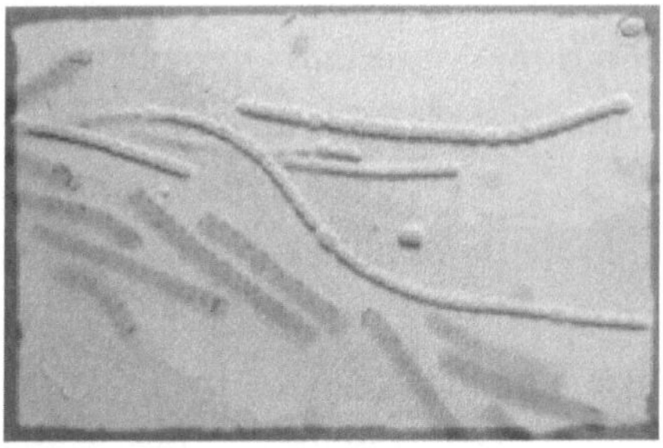

The vibration of the Earth as it cooled and the vibration of the first molecules on the Earth combined over time to create the next highest state of Consciousness - the Feedback Loop.

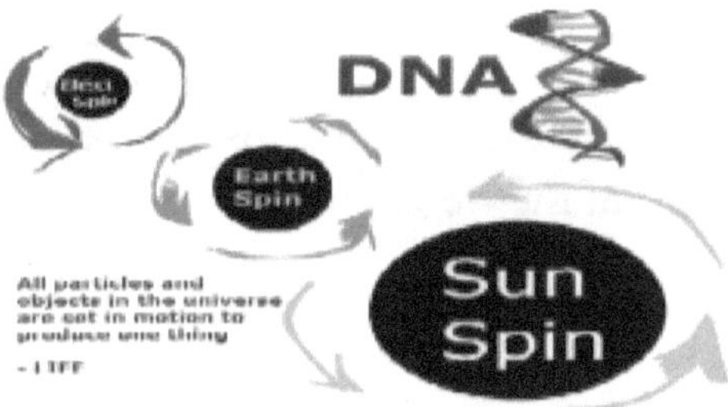

It's all about the Spin that we're in. It's the earliest spin gained from the God Particle Field that is handed down over countless generations and iterations of Consciousness.

Electrons spin, and this spin induces the spin of the atoms and molecules of which everything in the universe is composed. The spin of the Electrons is replicated in the spin of the molecules, Rotation, and from there, the Vibration of molecules to one another. All of this spin in the universe eventually leads up to all of human behavior.

The spin of the events in your own life is simply that - spin. Spin is not only a physical property of physics, it's also the way that things are interpreted by the code of life. From this point on, you should think of your life and the lives of others as spin. One life form spinning around another, or hundreds or even millions of lives spinning around each other, creates our history and evolution.

The spin we're in is deeply and eternally embedded in the code of life itself.

Flowers following the path of the sun across the sky are the best and most obvious example of the vibrational state of Consciousness that I can think of.

Bees are tuned to the Vibrational state of Consciousness too. This very simple form of life on Earth can see and feel the preferred wavelengths of light that the flower emanates, reflected from the sunlight to the bees eyes. He knows where the best chances for finding nectar are located and he wastes no time flying to it and carrying the precious energy back to his hive.

The Vibrational State of the universe makes this all very easy for the bee to stay focused on the flower because only certain wavelengths of light are offered up to the bee by the flower, depending on which variety of flower and which species of bees it needs to attract in this well-known symbiotic relationship.

Not only that, but the plant itself will change the color and pattern of its flower to something that appears like a 'Landing Field' to the bee in order to attract the right species to it. How does it know what the bee prefers to see and and how does it know how to do this? Answer - (There can only be one.) Consciousness.

As we all know, the white light from the sun is actually composed of all the colors of the rainbow as delivered by the varying wave lengths of light.

Flowers focus on giving back only one color each. How did they evolve in that way if they were not able to tune themselves to the Vibrational state of consciousness, as produced in the varying wave lengths of the universe.

Obviously, higher life forms will exhibit a more complex relationship to the vibrations all around us, but it's still a relationship to the Vibrational state that makes us what we are today. BUT this is just the beginning. To reach our fullest potential we must all reach the fourth and highest state of Consciousness - The Feedback Loop. Hopefully, every one of my readers will now know more about how to reach that state and how to be there more often and use it more fully to their distinct advantage.

Finally, I leave you with one more example of the Vibrational State of Consciousness that will make it clear why it is such an important state of Consciousness.

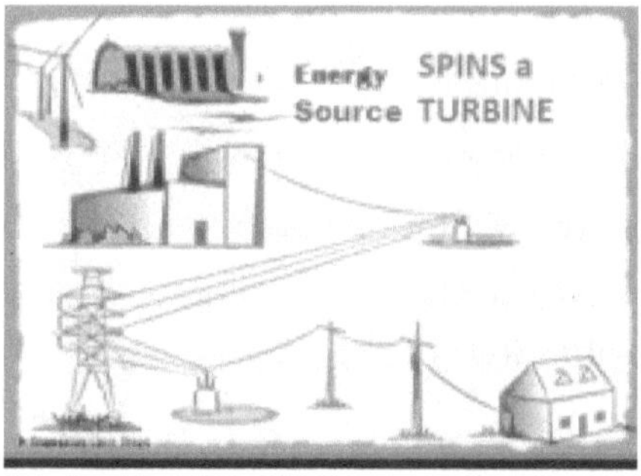

Whenever you turn on a light bulb or a computer or a hair dryer or any other electrical device in your home, the power plant that supplies your area with this most important energy has to know where and how much to send to power that device.

This is accomplished in the vibration of the switch going on in your device that travels back along the wires and all the way into the power plant itself where just that amount of electricity - say 100 watts for that light bulb or 200 watts for your computer are then sent back along the wires and into your home, in such a short amount of time that we never notice the delay.

The vibration of your device turning on is felt instantly by the power plant and it responds with just enough electronic power to make it work. This never fails unless there is some catastrophic failure in your area known as a black out.

But, the Vibrational state of Consciousness doesn't stop there. If your power plant is creating electricity by Wind Power, the vibration of you turning on your computer is felt all the way into the local wind stream that must now pour enough energy into your local wind generators to force them to turn just a little bit faster, just enough to power that extra device that you just turned on and demanded from the universe that it must work at that moment for you.

If your local power company is creating electricity from mainly solar panels, the vibration from your switch flows all the way from your home into the power plant and then back into the solar panels and from there into the sunlight that must now pour enough photons into the solar panels that will be converted into enough wattage again to power your device.

The Vibrational State of Consciousness is all about taking consciousness to the next level. The Vibrations of life and the rest of the universe are there for a more sensitive type of being to develop the wherewithal to respond to the vibrations of the universe and then give something back to it.

The more we experience the Vibrational State of Consciousness, in whatever art form we prefer, the faster and more thoroughly we can experience the fourth and final state - the Feedback Loop . . .

Because the universe is really just composed of a faster, richer and more powerful kind of music.

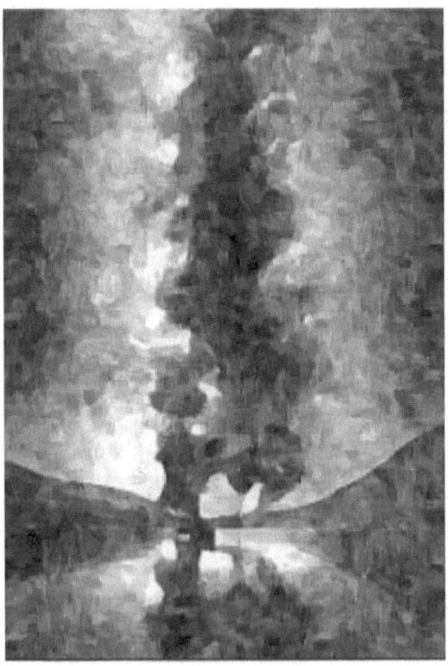

Chapter Four

- The Feedback Loop

Much of this chapter will be based on my earlier work – The God Particle Bible- where once again, you can find more details surrounding the subject of this book and you may also learn how I developed this theory of the 4 states of Consciousness after a lifetime of careful and considered deliberation.

The discovery of the God Particle in 2012 really got me thinking.

I call the 4th and highest state of Consciousness the Feedback Loop - because it seems to me that all of life was created out of this extremely beautiful process of induction - the will of life to exist out of nothingness - and that purpose is to give feedback to the Creator of all things. Because it is only through the feedback of our senses - how we see things, feel things, hear things, touch things, smell things - that our Creator can know how to improve upon all things. This idea is supported by the knowledge that God is Love. If this is true, then hearing from us, knowing about our response to this magnificent Creation is the greatest artifact of that Love. Love cannot go one way. For Love to be at its highest and best form, Love must be mutual. It has to be a two-way communication, a bilateral understanding.

It is important to note that all Humanity is at the point in our Evolution where our Feedback about our situation in the universe is louder and clearer and more profound than ever before. We are now able to take many of these lower forces of our Consciousness and use them to our greatest advantage, even to the point of knowledge where we can now change our DNA, thus changing our destiny in very major ways, not surrendering to the random events of Evolution and changes as has been the case for billions of years. This is a true "Sea Change".

We are placed at the highest state of Consciousness because it appears to me that all of the lower states of Consciousness are reaching for something that would be the ultimate goal of Consciousness - to know that someone out there is watching and reporting back, but not only reporting back, but also deeply appreciating what they find. This is partly from the Vibrational State as discussed above, but is fully reached at the Feedback Loop. With every moment that we appreciate our lives, the greater our connection to the Universe and the larger our Consciousness becomes.

Just as life is always striving to improve to the next and most intelligent state of a life form, the lower states of Consciousness are always striving to find the ultimate state of awareness. So, this is how the lower states are always rising to the highest state that they can find.

For this to happen, there has to be an observer, a final state of Consciousness that can see and learn about all of the others that make up the universe and report it back to the source simply because it is all about Consciousness and this kind of existence must also be known to itself, otherwise it is not true Consciousness, but merely accidental waves of harmonic energies.

Therefore, whenever any life form becomes aware of its own evolution or Creation, you begin this Feedback Loop. When humans first began wondering about the workings of the universe some thousands of years ago, we earned this highest rank of Consciousness.

In other parts of this galaxy, and all other galaxies, I am confident there are other life forms who also wonder about their origins and the inner workings of the universe and so they too have earned the 4th state of Consciousness and are probably currently writing books like this or making movies and/or documentaries like this one and the millions of other creative works by the millions of others like me all around our planet and theirs.

When we observe the universe using the level of technology that we have arrived at today, however, we alter the universe in some very subtle but very important ways.

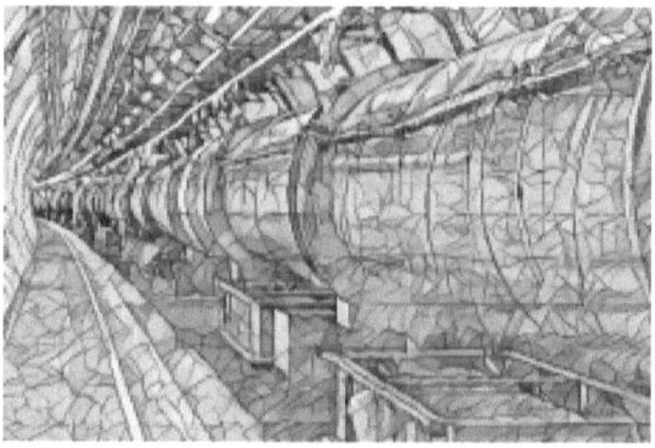

In the Large Hadron Collider, we recently discovered the God Particle, the Holy Grail of Quantum Physics because it was the single greatest missing link in our understanding of the modern model of Physics – out of what does mass form?

# # #

It is really the God Particle that starts everything in the universe on its march to a higher and higher Consciousness. I would include the God Particle in the Electronic State of Consciousness, only because there is no evidence yet of any Consciousness found at anything smaller than the Electron. However, having authored the God Particle Bible - there will be no one more excited than me if and when they discover that the God Particle itself displays any properties of Consciousness as the Electron does.

At that point, we may have to amend our theory to include the 5 States of Consciousness. At present I'm good with these four.

(Prepare yourselves for 'Zentanglements - The Three Laws Of Consciousness For Smarties coming in a more recent book.)

At the present state of my thinking, I'm convinced that the God Particle emanates from some conceptual part of the universe that is beyond Consciousness since it is difficult for me to think that Consciousness was created out of itself. There must be something even bigger and wider in scope and ability to achieve and create in the

UBER Universe somewhere that is able to create something as wonderful as Consciousness and have us thinking that's all there is.

# # #

And one thing we already know is that when we observe down to the subatomic level, we alter reality just a little bit because we must deliver energy or receive energy from this level in order to assess the qualities of these smallest forces.

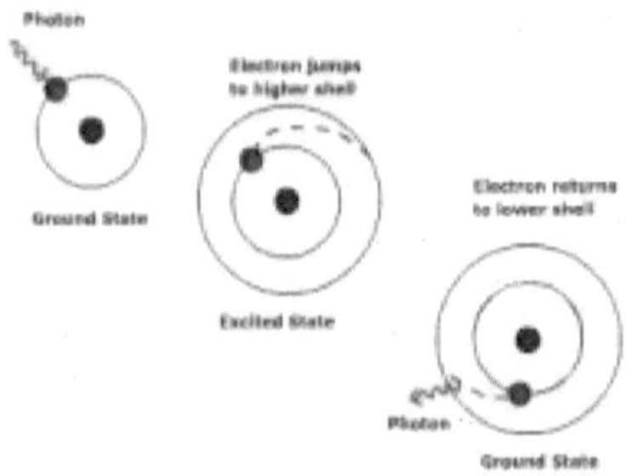

We cannot observe the electron with our eyes because to do so would send a photon of light onto the electron which is just enough energy to raise the Electron to the next orbital.

If an electron is induced to give up a photon of light, it will drop back down to the next lowest orbital.

So, when we observe things in Nature, we are either adding energy to it or taking energy from it. Thus, whenever we observe the universe we are raising the state of energy in the universe by an imperceptible amount - less than a drop in the ocean, far less - but by an amount of energy that we are putting into it.

When we learn from our observations, however, the universe gives up some of its energy in order to power our instrumentation and/or our brains and we therefore gain an energy state by our taking the next step in our knowledge.

We had to discover the power of Electricity in order to peer back down into the realm of the electron, the source of Electricity. This is no accident. It was planned long ago.

This is the nature of the Feedback Loop. Intelligent life evolves because the earlier forms of Consciousness must have our knowledge to learn about its own levels of Consciousness. Thus these lower levels of Consciousness give up the energy that is required in order for us to do so.

This is why when we learn something, we get an actual charge in electrical energy induced into our brains. Hopefully, you are feeling some of this energy right now. The brain being an electro-chemical process then vibrates with this new knowledge and sends this flow of energy back down to the original source of everything, the Electronic State of Consciousness. This completes the loop.

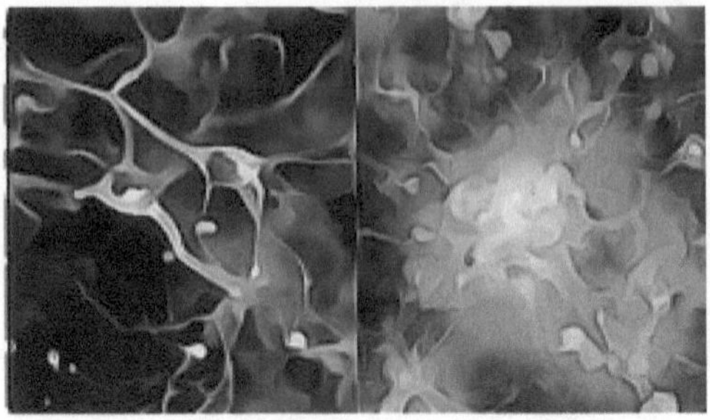

Left: The neuron net of the brain.
Right: The synaptic and galactic connections of the Cosmos.

It all started from nothingness. Electrons are born to balance out the proton energy inside the nucleus of the atom. Where protons and nuclei of atoms come from is also out of nothingness. Once these basic energies begin to spin into life, the Electron must come out of nowhere also just to keep the universe in an overall equilibrium. All of the positive energy in the universe, for example the positive charge of the proton, is balanced out in its entirety by the negative energy, in the subatomic world, the negative charge of the electron. But this happens everywhere else in the universe. You have only to look at all the aspects of your life to know this is true.

The net sum of all the force and power of the universe is ZERO.

As each state of Consciousness of the universe continually works to induce the next highest state, it must use up a great amount of power from the universe in order to gain that next step in its quest. When we observe these things, we give back part of ourselves. We give this energy to the universe in the required amounts to power the overall raise of consciousness in the universe. But, this lowers us by the same amount.

The Feedback Loop - all of life everywhere in observance of the events in their lives, the events outside of their lives - balances out the energy increase by giving back much of our own to the universe. But, in our learning process we gain the same amount back from the universe and this keeps the basic accounting back to a Zero Sum. Everything is in balance.

On the face of this news, you may think it impossible that life in the universe can contain enough energy to power the universe's continued evolution of Consciousness, but the point here is not that we contain enough power to power the entire universe's mechanics. We do contain enough power to maintain the climb from almost zero Consciousness to that which we obtain.

The power of Consciousness is not measured in electron volts or joules of energy or even in horsepower, but in something else. And, the power of one individual is almost infinite in the sense of its power to know things. We need only look at one or two lives in our history to see examples of this.

Albert Einstein, for example, one individual consciousness who could prove to the world how energy and matter are interchangeable to the degree that the atomic bomb could be created which we are now both blessed and cursed by today.

Isaac Newton who could put a 100% accurate mathematical measurement of the force of Gravity, as one other example. And this is how we land astronauts on other planets, not an insignificant achievement.

In the arts, Michael Angelo who could give us the greatest sculpture and paintings of all time or Leonardo DaVinci who could give us the Mona Lisa.

The Wright brothers who would learn how humans could fly. Thomas Edison who could give us light bulbs. Michael Faraday who would give us radio. Tesla who would give us the modern electrical power plant. Turing who would invent the computer.

The list of individual Consciousness is almost limitless and comes from one planet alone. I imagine that the list of planets where life exists is probably endless itself, all of it part of the Feedback Loop.

And in one location in the universe, here on planet Earth, the Feedback Loop has even learned how to decode our own molecule that is the code to our very existence. This single event closes the loop of our

Evolution where we can discover and work on the secrets of our own existence.

Everything, every minute, every day, every year and every hour in our lives is determined by this wave energy.

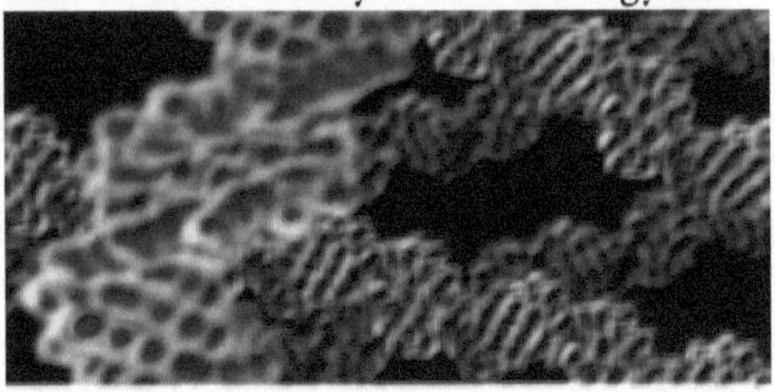

Whether or not we will have two arms, two legs, two eyes, two ears, one nose, one beating heart, how the heart will beat and how many times before we expire - all of this and all of the events in between are determined by our code of life implanted inside this molecule, which itself is implanted inside the nucleus of every cell of our bodies. All of the energies in the universe must also include all of the energies that powers up all of our own individual thoughts and actions.

This energy inside the code of life is derived from its component parts - the electrons and molecular energies we've discussed above.

A DNA cloud formation in the Sky as seen above my roof.

The God Particle comes first and puts God's spin, the very nature of Consciousness into everything.

We know this because the first levels of mass also have Consciousness. Two electrons when paired remember each other and know the state of the other.

When one electron spins UP, the other paired electron will spin DOWN. When one changes its spin, the other one knows and reacts to spin in the OPPOSITE direction to its partner instantly.

Since we are all constructed mostly of electrons, it is safe to assume that this state of Consciousness is passed all the way on up to our own level of existence. This is where we established the notion of 'Soul' and 'Soulmates'. If you're one of the lucky few on this planet, you may have even found your soul mate. If the condition of 'Soul-mates' did not exist at the subatomic level, we would not know this most wonderful and the most desirable aspect of human existence ever.

The most basic form of Consciousness that we have detected to date.

The Spin of the Electron and probably the spin of all particles inside the nucleus induces the next state of Consciousness - the Rotational. And so is the Spin of the Atoms that brings about the Vibrational or the Molecular Building blocks of our existence. It's all about the SPIN that we're in.

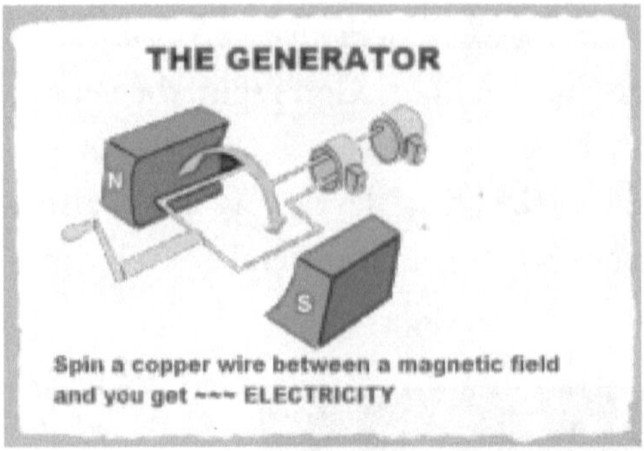

We also use spin to power up all of our devices, including the most advanced scientific instruments to look back down into the smallest particles in the universe and up into the biggest parts of the universe as well.

The Rotational State induces the next highest state, the Vibrational - which in turn induces the 4th and highest state - us, all life everywhere and what I call the Feedback Loop.

You and ME and all other forms of life in the universe are created by this master craftsman's machinery that started with the Big Bang and which continues to hum along today creating more and more life forms in myriad varieties everywhere in the universe at every moment in time.

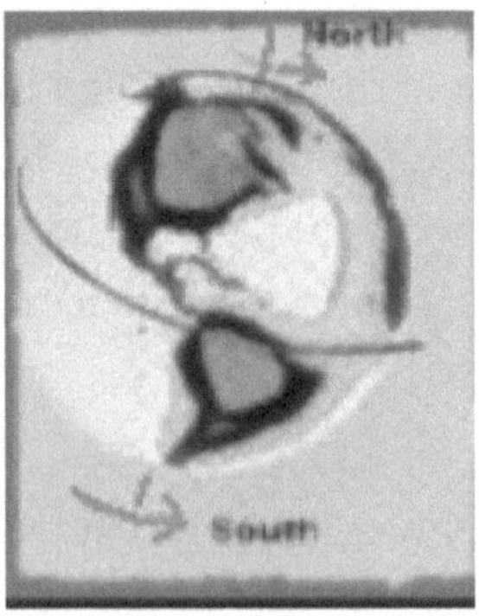

It's really just the energetic spin of things that creates electricity. And, on the Earth, it's the same thing - an energetic 'Spin' that created life on this planet.

The most basic form of Consciousness gradually creates life on the Earth by inducing the separate states of Consciousness until there is a final one - the one that can report back to the universe all of the wonder that brought life to exist in the first place. And, thanks to our highly advanced Science, this is exactly what's going on in our world today and is the most important part of our civilization.

This is our greatest hour, our zenith point in our evolution - to know that our major role in the universe is that of the Feedback Loop, the be-all and end-all of Creation. Without our powers of observation, we are nothing more than the dust and the fluids that make our molecular structure work together so beautifully in unison. Our powers of observation have never been greater or more refined or more tuned in to the vibrations of the universe. If we fail to live in this moment to the maximum best of our abilities and powers of the mind and heart - we are not worthy of the greatest opportunity ever given to any life form anywhere.

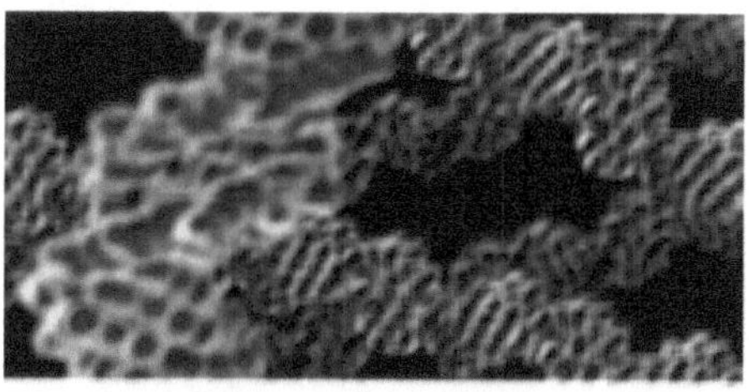

But all of this is already known to every one of you. Every last bit of what I have shown you is stored in the molecule of DNA that quietly sits inside the nucleus of every cell in your body. This is your most basic knowledge and is the formative energy of what we call instinct and intuition - our Extra Sensory Perceptions. You cannot avoid this fate even if you wanted to - but who would want to shy away from their responsibilities and the distinct honor of being the greatest and highest point in the universe - the reason it exists?

Not me.

Summary

- Don't Over-Think Consciousness

The ultimate highest and best understanding of the universe we live in is to understand how and why when we observe the universe we alter it. The energy that we use to observe the universe is the most important element in the universe because it produces the Art and Science that we all know and love and are inspired by when we know we have discovered another little sliver of Truth.

It is our duty to continue to observe the universe because when we avert our attention to other things, the mundanity of our everyday lives, the coming and going of the Kardashians, for example, we are allowing the universe to do whatever it will as expressed by the natural Laws of Physics. This means it will continue its random path throughout our communities.

When we observe the workings of the universe with every ounce of our being, with every thread of knowledge that we possess, we actually shape the universe around us in exactly the same way that we might shape a lump of clay. Our own reality unfolds in front of us as we observe it and make tiny little course corrections, again, based on what we know.

All of our lives matter. The more careful our observations, the more the universe will respond with equal care for our own personal relationship to everything else. Will your life improve based on your observations? Will the improvement be noticeable immediately or does it take a very long time?

I can't answer these questions for you. I can only say that this has been the blueprint for my life and is proven theory in my own mind because every single time I have noticed something significant about other people, other relationships, other events outside of my own, the more I get outside of my own skin and become conscious in a deep and profound way as to the affairs of others - my life is rewarded in very big ways. The facts that I've written down in this book are a fine example of the rewards that I have received for my keenest observations. I have no formal training in Science. I simply gained this knowledge from out of nowhere.

Every single human being - like it or not - is part of the feedback loop, the highest and best part of the universe. We have the evidence now that is the final proof that the universe was constructed to give us this ultimate power over it. We are composed of the natural products of consciousness itself, the God Particles, the electrons, the molecules, the nerves of our muscles, the neurons of our minds and thoughts, everything that composes us holds the properties of consciousness and these are enhanced and amplified at each of the 4 states until our own form of consciousness reigns supreme and brings it all together in a way that is unique. Our entire existence was formed for this purpose - giving feedback to the Creative forces that made us and everything else come into being.

The point of this exploration of mine in this book is that at this point in our evolution, we have already uncovered the major mechanisms wherein the universe is set up at the most basic and fundamental levels. We know the 4 States of Consciousness now and how these layers of the onion impact us all and make us think the way we think and do the things we do - all of it. Using this awareness in our observations of our future, we can alter it - greatly. We can make the greatest changes in the shortest time segments ever known.

Prior to these most recent and most significant discoveries about our place in the universe, it would take millions of years to effect truly positive changes in our communities. Today, we can accomplish more in just a few short months and years than all of the natural phenomena that would have taken millennia.

It's all in the power of our observations. As I write these words, on my TV set, a woman and her dog are rescued from the flood-waters around Baton Rouge, Louisiana. A man jumps in to pull a woman out of her car as it sinks into the dark muddy waters. Another man is standing on the banks of the floodwater taking in the scene with his video camera in his phone. The woman appears and the first thing she sputters is to please save her dog. The man dives down into the car one more time as it is slipping further into the murkiness and pulls her dog out of the car.

The video is shown to the world on TV and I observe this scene taking place in my universe and wonder why the man taking the video didn't abandon his role, jump in the water and help the man who was making this rescue.

And the answer is that we had to have - and we must always have - the observations of our world's events in order for them to have their greatest impacts. It was the observer taking this video who impacted the outcome of this event. The man who dived into the cold muddy water to rescue a complete stranger and her dog, may not have done it, except for the presence of the man taking the video and observing these events. The heroic young man who made the rescue was influenced to act in this heroic way and he may not have done so without the presence of the video camera monitoring these events.

Again as I write these words, a Youtube video is shown to me where two lions, a huge adult male and an adult female come across an injured baby fox. The poor fox is laying there in the dirt and probably knows that he will be a mere snack for these two huge and impressive wild animals. Instead of eating the tiny fox, the female first and then the male begin to lick the little creature's wounds. He begins to become more responsive and begins to move. The two lions encourage the injured fox to stand up. They push him with their tongues and noses forcing him to ignore the pain and slowly move around. After what takes almost an hour of nursing this grateful little animal, he walks away slowly, looks back at his rescuers and disappears into the jungle.

Again, the question in my mind is - would the two lions have acted to save this tiny creature if the man with the video camera had not been present observing this scene? Or would they have done what lions do and killed the little fox and eaten him?

The act of observing the universe - as in these real life examples - is the most important aspect of our lives because without it - we don't have even a clue about what we are supposed to do with the energy bursting forth from the interconnected machinations our bodily parts.

Without this aspect of our consciousness you and I are nothing more than recycled electrons and other space junk. The electrons that compose this universe have achieved immortality. They never die. They have also been gifted with the most basic and cleanest form of Consciousness. From this basic electricity, constantly charging up the rest of the universe to compose itself into the molecular structure of our minds and bodies, our own higher level of Consciousness has appeared. It is no coincidence that just as the universe is expanding constantly and the rate of expansion is accelerating, so is our own level of Consciousness right here on planet 'Urth'.

I hope that by now you have picked up on the fact that you and I and everyone else that you know are not really the person they portray themselves to be. We are all really just the physical embodiment of all of these states of Consciousness rolled up into one big huge glob of flesh and bone.

Inside of us is contained the Electronic State of Consciousness as imposed upon us by the existence of the trillions upon trillions times trillions of electrons in our bodies. But, we're also composed of the molecular biology that these electrons build up to. Then, from this Rotational State of Consciousness comes the Vibrational and then we're finally experiencing life as part of the Feedback Loop - the 4th and highest state of Consciousness.

Since we are all fairly vain creatures we may like to think that we live in the 4th and highest state of Consciousness all or at least most of the time. However, this is not true. We are products of all of the states of Consciousness - the sum and substance of all of the forces I've mentioned so far. Combined within us, they produce some of the most beautiful music, paintings, sculpture, architecture, roadways, telescopes and even space ships all from which we can project more of ourselves and learn even more from these forays into the rest of the universe.

And, to better illustrate the power and importance of the source of our inherited Consciousness - I leave you with a story that I just read about today.

A beloved husband and wife, who together raised six children, died hours apart in the same Alabama hospital in separate health-related incidents. A beloved husband and wife who together raised six children, died hours apart in the same Alabama hospital in separate health-related incidents, relatives said.

Jennifer Norsworthy, 40, was rushed to the hospital after a blood clot formed in her sleep last Friday. Norsworthy, who had a history with the condition, could not be saved.

Her husband, 38-year-old Toby Norsworthy, died les than 48 hours later on Sunday of a heart attack, loved ones said on a GoFundMe page created to offset medical costs.

Toby's sister, Tabitha Washburn, told AOL that Toby likely died because of the stress of losing the love of his life.

In a Facebook photo of the couple posted by Jennifer, she wrote, "A wonderful evening out with my hubby after a great day....could not love this man more!!! —feeling blessed.

The pair had been together for more than 13 years after meeting in Alaska and then moving to Tennessee, before finally settling down in Alabama.

Together, they raised six children, including three from Jennifer's previous marriage: Quinten, 20; Riley, 17; and Bradley, 13; and their children together: Mickey, 11; Aurora, 9; and 6-year-old Lainie.

"They both had a strong faith in God, were vital members in their communities and were such kind and gentle parents." Reads their Go Fund Me Page.

If this story - an amazing representation of the 4th and highest state of Consciousness that resembles so much the workings of the first state does not make you feel so lucky to be alive and wandering around on this planet here today - then none of this book will mean anything to you.

If, on the other hand, you feel something in your heart and soul for these complete and total strangers - then there is the last and final proof you need that these states all exist together today in perfect harmony and all we have to do is recognize them and how they work in our daily lives to truly appreciate our importance in space and time.

And one FINAL NOTE:

Some of my readers may actually still be wondering what this is all about and how this new information might actually impact their lives.

To these people, I would like to invite you, including those who believe they fully understand the implications of this research, to simply begin putting the information to use in your own lives.

Here's how to do it.

Make an objective and comprehensive look at your own life's events. Where are they taking you? Are you in agreement with where they are taking you? If so, how to put this information to use in furthering that direction. If your life's direction is not where you had intended in your earlier more idealistic times of your life, how can you put this information to use to change into the right direction?

Make a list of everything you want and everything you fear. What is the difference? How do they compete against each other in your mind's eye? How much time do you spend processing the thoughts that support them?

Next take a look at the direction that our civilization is taking and ask yourself if any of the events in your life are impacted by national or even world events and if so are these impacts in accordance with what you want and need or do they increase the likelihood of future events you fear?

Next, try to find the connections or entanglements between the events taking shape in your own life both good and bad and the events taking shape in the world as a whole.

Next try to classify each of the events into the state of Consciousness that best describes them. Are they best described and/or contained by the Electronic, Rotational, Vibrational or Feedback Loop states of Consciousness?

Make all necessary course corrections in your own life and ways to influence others to do the same in ensuring that total disaster does not come your way.

Lastly, remember always that the highest state of Consciousness in this universe is that of your own. There is no higher state of awareness other than that of God which is a state of Consciousness that is not capable of being defined by mortals such as myself, at least not objectively, with scientific proofs as I have tried to define the others here.

All of life on this planet and any other planet where life is found was created, is currently being created and will be created in the future for the sole purpose of giving feedback into the system so that it can evolve in a manner more suitable and propitious to life. We therefore are the be-all and the end-all of this Creation.

You must learn and appreciate more and more throughout your life how to use your feedback for this purpose - giving back the very best feedback that you can. For this effort, you are rewarded. Making the best effort you can in the realm of providing the highest and best feedback is your life's only goal that matters, one that will provide you everything you need from this moment on and into your ultimate future.

This is why I call the highest state of Consciousness - The Feedback Loop. You are part of it whether you like it or not. Your reactions to the events in your life are being recorded. You cannot escape this condition, nor would you want to. Everything you see, hear, taste, tough, smell or feel or think is part of the Feedback Loop. You cannot escape the nature of your existence. You are the sum and substance of your sensory organs. They were created for the purpose of giving feedback to the universe.

When you see something, the universe now knows what it looks like to YOU. When you smell something, the universe now records the information of those smells. When you touch something, the universe is now aware to another degree what something feels like to YOU. The point is, you're giving back to the source of your existence what it means to be alive and to observe all the things in the universe that you can and form a conclusion about each and every one of them, giving back to the Creator a modest amount of feedback. It's all appreciated and goes to the improvement of the overall experience – trust me.

When you hear your loved one's voice, when you listen to your favorite music, these sensations are broadcast out from your awareness in waves into the universe of all the other vibrations and they are absorbed and begin the process of altering the future in ways that may be more and more important to you. The level of importance to you is a direct reflection on the quality of your work.

When you truly appreciate the people, the places, the events in your life, you will begin to understand that none of them can harm you. They are all there to help you. Your reaction is what matters only. When you take on the world with this knowledge nothing can overwhelm your senses because they become trained and alert in ways that can defeat anything that may appear at first to be a harmful event. There is no harm coming your way, only life's constant flow of energy. It's all in how you perceive the energy hitting you and how you believe it will impact your future.

But the future is all inside of you as part of your fabric and you can do anything you want with this universal medium whenever you want. Given your new knowledge of how to take it all in, you now have the ability to deflect anything negative away from you and you will then make progress to anything positive that you may want to know.

Finally, I hope and pray above all else that every single one of my readers can see and understand that there is nothing difficult, nothing hidden, nothing not said, nothing that can be misconstrued in this knowledge. We have arrived at this point in our Evolution because of the constant workings of these 4 States of Consciousness over the last 13.7 billion years that we know about since The Big Bang. Everything was put in place by our Creator for the purpose that we may someday reach this level of awareness that gives us the insight into all awareness and how it benefits us. Why this universe was constituted in this manner, I leave up to you.

I believe I have given you all the evidence you need to draw your own conclusion as to who or what set up the universe in this most beautiful way that is a partnership, not a dictatorship. I think you may also begin to understand why it was setup this way after you have experimented a little with this knowledge and see it working in your own life.

The Final Test:

As the author of this book and the originator of this theory about the fundamental structure of the universe, I need your feedback.

In Science it has always been the keystone of advancing truth that the proof of any good scientific theory is for predictions to be made from the theory and then if other folks arrive at the predicted results from experimentation with the evidence as given here, then, and only then will you have a proof of the theory. If the predictions of experimentation can be duplicated by millions of my readers, then, there shall be no greater proof of any theory in history.

FOR PROOF:

Use the information you now have in your hands. Take on your greatest problem or challenge in life through the lens of this information. Try the following experiment. This has always worked for me. If it works for you, please let me know by putting your results on the Amazon Review page, found wherever you purchased the book and allow others to share in your results.

In regards to any problem you are having in life or any challenges you would like to take on:

1. Ask yourself: What are your electrons telling you?

2. What does the Rotational State of Consciousness know about this problem?

3. How does the Vibrational State of Consciousness shed any light on this problem or challenge?

4. What is your Feedback about this problem after looking at it from all of the first three states of your Consciousness?

In Step One, remember that your electronic state of Consciousness came first and therefore it is the oldest and wisest of all levels of Consciousness. The influence that billions upon billions of electrons have on your mind is a subtle one, but it is always there, egging you on, encouraging and discouraging you according to their own priorities. But, they are also very wise in the ways of the universe in greater proportion than you yourself, because the 'You' that you think is 'You' is really the summation of them. But, they love you and always want the best for you in the long run, therefore, you can trust their advice. You already know this, because they live inside you and have the most intimate symbiotic relationship with you that you will ever know. Find your quiet place and just listen to them.

In Step Two: Your electrons also have their own social needs, so they clump together to form atoms that will eventually form the molecules that are your flesh and bone. But, prior to forming, they have a rotational energy, a Consciousness that is beginning to flow in ways that are literally spinning in space just as the planets spin and revolve about the sun, and the sun revolves around the center of the Milky Way Galaxy and that all galaxies of stars everywhere revolve around the center of everything. All of this motion inside and outside of you gives you the overall perspective on life. What is this gravitational force field's effect on your situation?

In Step Three: In the Vibrational State of Consciousness, you want to know what your heart is telling you. How do you really feel about the problem or challenge in the deepest parts of your soul? Remember, we're talking about music here. Is there a musical connection that you can explore in your situation? If so, does your 'Music Appreciation' lead to a new kind of music? Feel the problem. Resonate with it. Don't fight it. Learn about it by grooving to the music it puts forth if any. Adding some of your own notes may make up a new band or create a new song for you. Live your life to the fullest and help others do the same - Enjoy!

In Step Four: The Feedback Loop, the highest state of Consciousness, is simply the acceptance of what you have gathered up in knowledge throughout your life and telling someone about it. In teaching truth to others, you learn the truth in its highest and best form. This book, for example, is my greatest expression of my own Feedback to the universe. It is being assimilated by you right now and also by the Source of the universe and I'm aware that the source of everything is very interested in the feedback that I've put forth here. I'm fulfilling my own destiny and there is no higher reward than that.

How do I know this? Because I am further encouraged by all of the events in my life, big or small every day. I noticed this from the time when I was a small boy. I nurtured it all throughout my life and noticed the good and bad things that happened to me in the most objective ways I could in order to learn about this process we call life. The greatest lesson to me, from the time I was a small boy is that someone or something was listening to my feedback, the results of what had been given to me during the day and my feelings about them.

But when I express my gratitude for what is given, I receive much more the next day, or the next minute, or the next month or year. So in every feedback I give to the universe there is an equal expression of love that comes back to me every time.

Try it out. Show your gratitude for what is given and you may find bigger and better things coming your way soon after. When you cannot be grateful for what is given to you, don't whine or complain. It does no good. Be grateful for the 'Bad' as well as the 'Good' and you will find the 'Bad' stuff gradually fading away. When 'Bad' energy has no negative impact on you, it diverts to other receivers of 'Bad', those who want it or don't know how to deflect it.

I call the Feedback Loop, the 4th and highest state of Consciousness because it is the connection to our Creator. The source of all your sadness or happiness comes from the energy that the Creator created in the form that he or she maintains throughout all of the Creation. The four states of Consciousness are akin to our streets and roads, the schools and rules that we all must follow in our life's path. There is no avoiding any of it, unless you want to lead a very uninteresting and unrewarded life.

Give the Creator as much feedback as you can muster and you will find your greatest challenges or problems completely overcome by your energy and in ways that you never dreamed possible.

Please conduct this experiment and I predict a million or perhaps even a billion happy lives as a result. And that will be my proof.

# # #

Go forth, live a long and loving life and prosper.

If you forget everything else I've taught you, remember, at least this one thing - You were made in the Spark of Life - The very FIRST form of Electronic Consciousness that we know, but far from the LAST.

About The Author

We want to see The Love Bot made into a major motion picture. If you agree and would like to see this story in theatres and/or streaming on your TV—-PLEASE GIVE THIS BOOK AN HONEST REVIEW from wherever you found my book.

It really helps the book gather more readers – to get the TRUTH about our tenuous future. – Thanks!

I like to describe myself as a life-long student of Science and Science Fiction writer with a specialty in the area of Consciousness. With a college degree in Psychology, I was underwhelmed as an undergraduate working in an area with a total lack of any real research into the question of how Consciousness makes us think about things the way we do? And where Consciousness might reside in our brains?

The personal computer revolution would start to yield some of this research though not from Academia, but out of Silicon Valley. As computers got smarter and smarter, they tended to force us to ask these questions until one day an artificial intelligence may emerge from a lab somewhere that can outthink even the smartest living individuals on the planet. Some say we've already reached this historic point in Evolution they call – The Singularity'. .

Being in the computer business and an early adopter of the Internet, I had plenty of time to think about this idea. I studied Genetics, Biology, Evolution, Computer Science, Economics, Quantum Physics, Astronomy, Crypto-Currencies and through it all I uncovered a pattern, a train of thought, a subtle yet living filament of Truth that had yet to be appreciated.

You can follow this filament of truth from the many books I have authored along my path. The first major epiphany of mine came from the announcement of the discovery of the God Particle or Higgs boson from research done at the Large Hadron Collider near Geneva, Switzerland. This single energy is now considered to be the cornerstone of all modern understanding of the forces in the universe. Until July 4th, 2012, when they made this momentous announcement, we had no idea about the real nature of the Big Bang.

In The God Particle Bible – I lay out my theory that the God Particles were sent out FIRST like an advance attacking force from the Big Bang because since these invisible eager beavers give every other particle their mass, they would have to be first ones out of the gate in order to create the first electrons, protons, neutrons, quarks, etc that eventually congealed into the stars, galaxies, planets and eventually even us.

And so I say - "It's no accident that these particles were nick-named - 'God Particles.'"

From here, I started to think about all of the other energies from the God Particles on up.

The Electron, for example, somehow knows where to orbit the nucleus of every atom it joins in order to create. There are only a handful of distances away from the nucleus where they have permission to orbit. And, every single atom in the universe, as far as we know, and of which there is no number large enough to count them - contains these same exact places where the electron energy can live and no others.

But these places have long been known. More recently, however, it has been shown that electrons can also pair up, or become 'Entangled' with another electron. From this point forward, each electron in the pair knows about and responds to the existence of the other and so the question I then asked myself is a simple one: 'How do they know?' They're far too small to have brains or any other kind of neural network, as far as we know.

From these questions came the book - 'The 4 States of Consciousness' wherein I demonstrate how Consciousness begins at the subatomic level and is passed up through three more levels of Consciousness and into our own existence.

Remember, your own brain is running on the Electronic forces that transmits a signal to billions of neurons inside your brain which result in ALL of your thoughts. The first state of Consciousness, is the Electronic state and so it is no coincidence that the first state ends up in your brain, the 4th and highest state of consciousness in the created things of the universe.

In a more recent book, 'The Science of Physics - Proof That God Exists', I go further and create a new number and even a simple formula based on my new number - that demonstrates how and even why the universe is constructed in this extremely intricate and intimate manner.

I want to be remembered as the fella who discovered the 4 States of Consciousness and the Three Laws of Consciousness.

And if I will be remembered for solving the world's entire energy crunch and all other social and political problems along the way – that's a bonus.

If you're interested in learning more about how this story came to be, you will find clues in some of my earlier books such as:

Tree-Quivalance – Orac Saves the world AGAIN

The A.I. Bible – Orac Saves The World

The Science Fiction Bundle – Combines The Jupiter Sun and The Book To End All Books.

The Jupiter Sun – Creating a 2nd Sun In the Sky
The Book To End All Books
My Daily Environmental Journal
My Cosmic Brain – The 2nd Big Bang
Zentanglements – The Three Laws of Physics for Smarties
The Science of Physics – Proof That God Exists
The God Particle Bible
The 4 States of Consciousness
The Origin of Creation
The Blockchain Government
Metamorphosis: # AI Cocoon
Extinction Live
Anti-Matter
Brain Drain
Maxtricity
Can Electrons Learn?
The God Particle Bible
. . . and several others -

All of which can be found online, other places where books are sold as well.

Just Prior to this book, I authored:

AI Chat GPT
Elections 2024
Orac For President
Michael Mathiesen

In this book, Orac actually runs for President of the United States in 2024.

Most voters today claim that they would like to see someone "Completely New" run for President this year. As you have now learned - You won't find anything much 'Newer' than Orac.

Just Prior to this book, I authored:

In this book, Orac helps discover the follow up to E=mc2

The Force (F)= meG

Just prior to this book, I authored:

Which combines the following three books into one combined set.

Prior to this book – I authored:

This is also collaborated by Orac my AI Chat Assistant and focuses on solving the problems we face as a country.

OR you can get any of these three books separately.

Prior to this book I authored:
Tree-Quivalence – Orac Saves The World - AGAIN

This book is also actually written by Orac, not me. I merely asked some penetrating questions. ORAC gives us a convenient way to think about all of our daily activities, how they are damaging to our planet and how to pay back to our mother so that she may live forever, giving life to this part of the universe in safety and harmony forever. Artificial Conscsiosness AC is demonstrated here not just artificial intelligence. AI.

Prior to this book, Orac and I collaborated on .

This book was also in collaboration with Orac in which he saves the world.

Prior to this book, I authored:

This is the Science Fiction Story that gave me the realization that we could indeed ignite Jupiter into a Sun In my mind, at least, it was no longer just fiction.

SO Just AFTER this book, I authored:

The Jupiter Sun

This book may become much more famous than it is right now because it proposes that we go out to Jupiter in our SpaceX and NASA missions and actually shoot enough laser energy into the core of Jupiter to start a Nuclear Fusion Reaction exactly the same as what happens inside our own sun and thereby create a 2nd Sun.

MichaelMathiesen.com[1]

OR GET THE TWO BOOKS that are all about The Jupiter Sun – The Science Fiction Story AND the real Science behind how to actually DO IT[2]

1. http://www.michaelmathiesen.com/

2. https://www.amazon.com/My-Best-Science-Fiction-Jupiter-ebook/dp/B0BSYJVF43/

Just prior to this book and the inspiration for The Jupiter Sun , I published

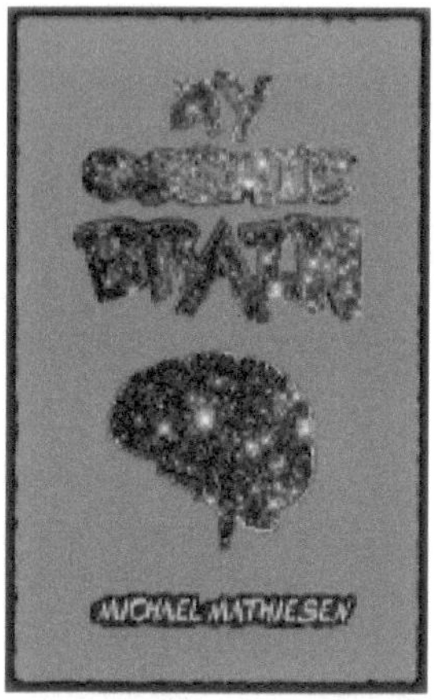

This is a wild story of another version of the near future as seen by someone who has donated his brain to science.

Prior to this book, I authored 'Zentanglements – The Three Laws of Consciousness for Smarties.

I also illustrate most of my books and covers. You can purchase most of the signed original artwork for this and all of my other books at:

CafePress.com/TheLoveBot[3]

Supports a Good Cause
Here's a few samples:

---

3. http://cafepress.com/thelovebot

This is called 'Buddha In The Backyard' because I found this statue of the Buddha sitting at the head of a parking space near to my house and loved the composition.

Here's another one of my creations:

This is called – 'DNA in the Sky'

On the day that I finished my book 'Zentanglements, The Three Laws of Consciousness for Smarties', I went out into my backyard to relax. I looked up at the sky and noticed the cloud formation above my house that reminded me of the DNA double-helix molecule that makes up every cell in our bodies. This was a huge affirmation for me because in an earlier book, I had put out the theory that the universe is actually traveling through the heavens in the form of the same DNA molecule. Well, I've never seen a cloud formation such as this before in my life and probably never will again. It turns out that many other things in the universe have this basic structure.

In much of my writing I like to tie in to the structure of the DNA molecule as part of other structures we see in Nature. It turns out that new organic solar Photo-Voltaic minerals that I talk about in this book are also Chiral in their molecular structure or Helical.

HERE'S ONE MORE AND, they can all be purchased at –

CafePress.com/TheLoveBot[4]

OR - CafePress.com/myfineart

4. **http://cafepress.com/thelovebot**

This is called the 'Hand of God' Nebula. Discovered by the Hubble Telescope a few years ago, it is one of my favorite images. It's actually out there in Space watching over us all.
Are we special? Does life on Earth serve a Cosmic purpose therefore must we guard it with our lives?
I graduated from the University of California at Santa Cruz with a B.A. In Psychology. I majored in Psych because I felt that somehow, someday I might be able to make a major contribution to that Science vs other areas where I would not be able to contribute much at all.

I believe this was a very wise choice of majors for me because since those wonderful days as a student, I have written over a hundred books about the mind, Consciousness, health, government, Scientific discoveries, all things that interest me and require a strong mind and profound curiosity to enjoy and explain them to others. AND, that's what I have done for most of my life. Would I change any of it in any way? – NOT A CHANCE.

I also studied Art and made my first painting when I was in an Art class and I enjoyed it immensely. I always like to boast that I sold my first painting, and several others later on which is many times more successful than Vincent Van Gogh who never sold a painting when he was alive.

In my book prior to this one, I combined my artistic talents with my observations of the universe and came up with a book about Consciousness and I even found myself able to define what I call –

The Three Laws of Consciousness.

They are:

Consciousness creates all things.

Consciousness connects all things.

Consciousness consists of all things.

I'm very proud of my three laws and I believe I should be entitled to the Nobel Prize in Physics for working this out because these laws, as you may read in the book are based upon the latest discoveries in Quantum Physics. I even invented my own formula to describe this area which is expressed thusly.

Zero to the power of Infinity equals any number in the universe. More can be found at my website: (Someday, this formula could become as powerful to us as E=mc2)

- Thank You For Listening -

PLEASE GIVE THIS BOOK AN HONEST REVIEW and post them to Facebook, Twitter, or wherever you found my book. It really helps the book gather more readers – listeners who need to know the TRUTH. – Thanks!

Prior to authoring this book – I authored:[5]

The Science of the Bizarro Universe and how to avoid it.[6]

---

5. http://udemy.com/zentanglements

Prior to authoring this book – I authored:[7]

The Blockchain Government[8]

You've heard about the blockchain and how difficult it is to hack it - while our present system is so fragile that many politicians can pretend that they won even when they didn't.

6. http://udemy.com/zentanglements

7. http://udemy.com/zentanglements

8. http://udemy.com/zentanglements

Then, as soon as they get into office, they forget all about US
Narrated Listening Versions available – MichaelMathiesen.com
We spend TRILLIONS of dollars per year for simple government services that could be accomplished for a few thousand bucks by using the Blockchain.
Prior to THIS I AUTHORED

All Versions [9]availalble
Extinction.Live [10]
What what happens LIVE as the Human race goes extinct.
If you liked The Jupiter Sun – you'll like this one.
NOT for the faint-hearted! But, could become reality if enough of us don't start making BIG CHANGES!!

This is a Science Fiction novel based on the notion that MARS may be our best hope to prevent the coming catastrophe that could spell the extinction of all life on Earth.[11]

9. http://www.extinction.live/

10. http://www.extinction.live/

MOST of my books can also be found as Narrated Listening Versions BOOKS wherever you found this book.

Prior to this book – I authored -

11. http://udemy.com/zentanglements

Narrated Listening Versions at M[12]ichaelMathiesen.com
This book is the one that should have won me the Nobel Prize in Physics by now. Probably after I'm dead. [13]
The science uncovered here will amaze you.
The more time goes by, the more I believe I hit the math on the button and that it proves this book is as accurate and predictive to the Human Race as Einstein's best work in his theory of General and Special Relativity.
Proof That God Exists? My New Mathematical formula proves Einstein's Unified Field Theory, how and where God exists and String Theory ALL at the same time.[14]

In one of my earliest books, I solve all of this country's political woes by detailing the things we must alter in our system to bring it up to the modern standards of technology and the Economy that we know today.-

America 2.0 - Take Stock In America
What if we incorporated the whole damn United States of America into ONE HUGE Social Benefit Corporation? We are all stock holders and we all vote our shares?

12. http://udemy.com/zentanglements

13. http://udemy.com/zentanglements

14. http://udemy.com/zentanglements

America 2.0 Inc.

Take Stock
In America!

Michael Mathiesen

Political pundits keep saying on TV that the United States is a DEMOCRACY – That's a LIE. In a real democracy as defined by the original Greek Inventors of it – the MAJORITY DECIDES any of the most important issues by drawing black rocks or white rocks. The color of rocks that gets the most – wins the day.

We need to INSTATE THAT HERE IN THESE UNITED STATES or continue to be dominated by Political Parties and Special Interests until the day they WIPE US ALL OFF THE MAP.

Then, I thought perhaps humans are just not meant to rule over themselves because they will always conceive of trickery to game the system no matter what. SO, I imagined a planet ruled by robots. I know – sounds awful – but what if they were not only smart – but very attractive and sexy?

Autonomous Government
Introducing Alterra
Michael Mathiesen
Sample

AUTONOMOUS
GOVERNMENT

Imagine if a robot like Alterra were the leader of the world and that there were thousands of copies of her that roamed the planet, without bodyguards and just listened to and then quickly solved all of our problems.

Prior to this book - I wrote:[15]

Much of the ideas for this book originate in this adaptation of what Representative Alexandria Ocassio-Cortez and Sen. Ed Markey have proposed as possible future legislation. WE HAVE TO WORK WITH THESE PROGRESSIVE LEADERS for this concept to have a [16]

chance. There are a few in Congress who know that the END IS NEAR if we don't completely change everything including how we earn our money every day.

UNFORTUNATELY they are in a tiny MINORITY – so cannot get it done in time to save the world – in my opinion unless something ELSE HAPPENS.

15. http://udemy.com/zentanglements

16. http://udemy.com/zentanglements

LIKE ENACTING A GLOBAL CONSTITUTION

And so, recognizing this crying need - I wrote.

I'm actually ANONYMOUS.

It has to happen someday – Why not **PRIOR** to when we kill our entire civilization?

Please go to where you found this book - search for these books and then find on the bottom of the sales page where you can add a few sentences and give it an honest review - Thank You.[17]

M[18]ichaelMathiesen.com

Please further your education by taking either the PRINT or Narrated Listening Versions of any of my titles. [19]

AND, is the bad stuff that happens to us really just Karma? Or is there a better scientific explanation?[20]

---

17. http://udemy.com/zentanglements

18. http://udemy.com/zentanglements

19. http://udemy.com/zentanglements

20. http://udemy.com/zentanglements

Give it a read. It's a fun and easy read if you like classical Science Fiction.

What if over 90% of the universe was made of Anti-Matter? It's not and no one really knows why.

Just prior to this book, I authored:

The 4 States of Consciousness[21]

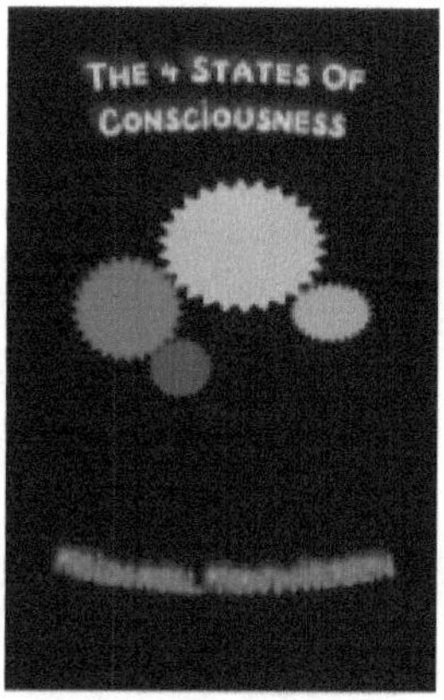

21. https://www.amazon.com/4-States-Consciousness-Michael-Mathiesen-ebook/dp/B01EZ7Y7X4/

Because the only way we have obtained our own level of Consciousness is through the gradual evolution of the lower 3 states of consciousness that induces our own, the 4th and final level of consciousness that we can measure..

ONE of my Earliest works that evolved into my life-long passion is in my earlier work. I consider this as the start of my 'Honest-To-God Scientific Journey'.

The God Particle Bible

This could be the 2nd Greatest Story
Ever Told

When the God Particle – (Higgs boson) was discovered by the Large Hadron Collider, I was blown away, and so was all prior thinking about how the universe is constructed and who or what constructed it. Get this book – TODAY!

cafepress.com/thelovebot[22]

For some of my original artwork, such as the cover to this book.

AND, most of my books can also be found as Narrated Listening Versions at my own website

MichaelMathiesen.[23]com

Finally, please order a copy or a number of copies of this book so that you can have extras in your library that you can then lend to other people.

AND You can find T-Shirts, Hoodies, and other clothing items with our logo printed on them.

LET THE WORLD KNOW YOU CARE!
Wear ORAC's HATS and T-SHIRTS, MUGS, APRONS, ETC. PROUDLY—

Cafepress.com/TheLoveBot[24]

22. http://cafepress.com/thelovebot

23. http://www.michaelmathiesen.now.site/

AND

cafepress.com/aichatgpt

LET THE WORLD KNOW

Thanks for reading.

AND – we want to see The Love Bot made into a major motion picture. If you agree and would like to see this story in theatres and/or streaming on your TV—-PLEASE GIVE THIS BOOK AN HONEST REVIEW ON Facebook, Twitter or wherever you found my book.

~~ SHORT REVEIWS REALLY HELP ~~

Can I count on YOU?

It really helps the book gather more readers – listeners who need to learn the TRUTH. – Thanks!

For related ideas please visit my website:

www.LoveBotMovie.com

And if you can please donate money to One Tree Planted – a non-profit dedicated to planting a tree somewhere in the world at a cost of only

ONE DOLLAR PER TREE PLANTED

Probably the best investment you will ever make.

Even ONE DOLLAR makes a difference.

OneTreePlanted.Org[25]

Thank You

---

24. http://cafepress.com/thelovebot

25. https://www.onetreeplanted.org/

One Tree Planted can compensate for the amount of environmental damage that all of us do to our planet every day just by living our lives. One Tree Planted gets a tree planted somewhere in the world for only one dollar.

You won't find a better bargain than that. Sadly, we'll need trillions of trees to save ourselves – but One Tree Planted can and is reaching that goal for us. But they need you.

MichaelMathiesen.com[26]

26. https://www.michaelmathiesen.com/

www.ingramcontent.com/pod-product-compliance
Lightning Source LLC
LaVergne TN
LVHW090414210926
848480LV00002B/569